Electrical Safety Engineering of Renewable Energy Systems

Electrical Safety Engineering of Renewable Energy Systems

Rodolfo Araneo
University of Rome
Rome, Italy

Massimo Mitolo
Irvine Valley College
Irvine, California

Library of Congress Cataloging-in-Publication Data

Names: Araneo, Rodolfo, author. | Mitolo, Massimo A. G., author. | John
    Wiley & Sons, publisher.
Title: Electrical safety engineering of renewable energy systems / Rodolfo
    Araneo, Massimo Mitolo.
Description: Hoboken : John Wiley & Sons, 2022. | Includes bibliographical
    references and index.
Identifiers: LCCN 2021025391 (print) | LCCN 2021025392 (ebook) | ISBN
    9781119624981 (hardback) | ISBN 9781119624998 (pdf) | ISBN 9781119625018 (epub) |
    ISBN 9781119625056 (ebook)
Subjects: LCSH: Electric apparatus and appliances--Safety measures. |
    Renewable energy sources--Safety measures.
Classification: LCC TK152 .A73 2022 (print) | LCC TK152 (ebook) |
    DDC 621.3028/9--dc23
LC record available at https://lccn.loc.gov/2021025391
LC ebook record available at https://lccn.loc.gov/2021025392

Cover image: © oxygen/Getty
Cover design by Wiley

Contents

Preface

To my wife Stefania, for her unfailing love, patience, and encouragement,
and to my blessed sons Leonardo and Giulio, who give me the tremendous
energy of the youth.
To my wife Jennifer and my daughters Alessandra and Giorgia, who are
the true renewable energy source in the heart of their father.

It was the year 2004 when Prof. Araneo began his journey through renewable energies. And it was the year 1990 when Prof. Mitolo started his in electrical safety engineering.

Prof. Araneo was asked by a high school and college mate to jointly work on the design of one of the first 1 MWp utility-scale photovoltaic plants in Italy to be built in Sardinia. He was an inexperienced, fresh PhD graduate in electrical engineering, but driven by an innate curiosity and interest in all that was electrical engineering. He did accept the challenge, and in the years to come, he worked on the design and construction of almost 1 GW of renewable energy plants. In the early 2010s, he started teaching a university graduate course on renewable energies.

Prof. Mitolo, a fresh PhD graduate in electrical engineering, read the EU Directive 89/391/EEC, also known as the *Framework Directive*, which indicated that the *risk assessment* was to be the cornerstone in the prevention of accidents. This, right then, confirmed in his mind the idea that electrical safety could not only be implemented through a list of prudent actions near energized parts but mainly via the proper design: the *safety-by-design* approach to anticipate and design out hazards. Since then, Prof. Mitolo has accordingly designed, researched, taught, and authored.

These authors met during the Christmas Holidays of 2018 in beautiful California and spent time enjoying with their families the beautiful scenery of the Pacific Ocean, and, at the same time, discussing the typical academic issues (i.e., conferences, papers, research, and grants – topics that their families found boring, especially during those sunny days). It was then that Prof. Araneo forwarded the idea to write a book on renewable energies and safety. Some may decide to write engineering books after long and erudite meditations; others (i.e., these authors) did it under California's blue skies, surrounded by family.

An amazing (and demanding) period of two years of work and dedication started, whose results the authors hope the readers will enjoy.

The authors believe that a book is the ideal means to make it possible to share the results of experience and research with the readers. Many books on renewable energies have been written, yet this book introduces a different view on the topic: it teaches the practitioner (student or engineer) how to ensure electrical safety in renewable energy systems (e.g., when the sun shines, a PV array cannot be turned off).

This book does not want to be the classic academic book for a university course, but it has been written with the intent to marry the experience of the professional engineer with the scientific rigor of the university professor: two ugly beasts that together may have produced an interesting result.

Nowadays, there is a great and unsatisfied demand for electrical engineers, especially in power systems, which now do include renewable energies. This is especially true in light of the global energy transition that we are experiencing: we are walking along a pathway toward a transformation of the global energy sector from fossil-based to zero-carbon footprint. The electric engineer will play a major role since safe electrification is expected to be a critical milestone for decarbonization. There should be no shortage of challenging problems in renewables in the foreseeable future. There will be a clear need for engineers to have both an understanding of the fundamentals of electric safety as applicable to renewables and the creativity to apply this knowledge to problems of practical interest.

This book is for those engineers.

Acknowledgments

It is a pleasure to acknowledge the role of many students in helpful discussions. Prof. Araneo would like to thank Giorgio Mingoli for the phone call in 2004. The authors would like to thank the following persons for providing materials and being available: Giuseppe Mastropieri, Fabio Amico and Massimiliano D'Angelo of REA Advisors, Raffaella De Cupis and Gianfranco De Simone of Ghella, Massimilano Donati of VEI Green, Saverio Spampanato of Secundm Naturam, and Riccardo Cecere of Enel Global Power Generation. Finally, Prof. Araneo finds it difficult to express in words his gratefulness to Prof. Salvatore Celozzi, who provided him with guidance, support, and encouragement during his career.

Prof. Rodolfo Araneo
Roma, Italy, March 28, 2021

Prof. Massimo Mitolo
Irvine, California, March 28, 2021

1

Fundamental Concepts of Electrical Safety Engineering

> *I shall be telling this with a sigh*
> *Somewhere ages and ages hence:*
> *Two roads diverged in a wood, and I-*
> *I took the one less traveled by,*
> *And that has made all the difference.*
>
> *Robert Frost*

1.1 Introduction

The renewable energy sector has been rapidly growing in the past decade [1–3], and so has been the number of accidents involving workers in "green" projects. Statistics in the United States reveal that injuries and death are caused by

Electrical Safety Engineering of Renewable Energy Systems. First Edition. Rodolfo Araneo and Massimo Mitolo.
© 2022 by The Institute of Electrical and Electronics Engineers, Inc. Published 2022 by John Wiley & Sons, Inc.

lack of safety training and safety procedures [4]. The *Electric* hazard, but also *Falls*, *Struck by* and *Caught in between* hazards, are always present during all photovoltaic, solar thermal, and wind tower construction projects, regardless of the magnitude of the job.

The culture of the *safety-by-design* [5, 6] seems to be the appropriate response to the increased risk offered by renewable energy systems (RES). RES may challenge the safety of workers because they are generally always *live*, and the system voltage may exceed 500 V d.c.[1]

In addition to safety training and procedures, electrical safety may be conveyed through engineering measures that reduce the risk of electric shock below a threshold that is conventionally deemed acceptable by applicable standards. In fault-free conditions, the *basic* protection ensures that persons cannot come into contact with parts normally live (i.e., proper insulation of electrical components). In the case of failure of the basic insulation of components, the *fault* protection ensures defense against electric shock by automatic interruption of the fault current. In some scenarios, the fault protection may be obtained with alternative methods to the fault current interruption.

In general, the *safety-by-design* of RES [7] is achieved if hazardous energized parts are never accessible, and that equipment/appliances, also referred to as *exposed-conductive-parts* (ECPs), are never hazardous either under normal operations or in the event of single-faults. In essence, touch voltages and contact durations must be within the magnitudes deemed safe by applicable technical standards and codes.

1.2 Electric Shock

External electrical stimuli applied to the human body can prevent operational skeletal and cardiac muscles from properly operating, as well as destroy bodily tissues by thermal shock.[8]

External a.c. currents with frequency ranging from 50 to 100 Hz of magnitude around 10 mA for adult males and 15 mA for adult females, can override the internal electrical signals from the brain controlling the body muscles, render the person unable to "let go" of an energized part and cause painful muscle contractions.

For d.c. currents, thresholds of let-go cannot be positively defined. The circulation of d.c. current through the body only causes a sensation of warmth, and the person is subjected to painful muscle contractions only during making and breaking of the d.c. current.

---

1 The system voltage of PV arrays is defined as $1.25V_{OC}$, where V_{OC} is the open-circuit voltage of the array.

Stevens' Law [9, 10] describes the perceived strength of a physical stimulus as a function of its intensity, expressed in its physical units. According to Stevens' Law, the perception of electric shock is *superlinear* with the stimulus, as varies as the 3.5 power of the a.c. voltage applied. The coefficient 3.5 relates the magnitude of the applied voltage to the perceived magnitude of the shock as current through the fingers; thus, a small increase in the applied voltage is perceived as a larger increase in the electric shock.

A 30 mA-current, if interrupted within 300 ms, can cause involuntary muscular contractions but usually no harmful electrical physiological effects. Longer disconnection time, up to 5 s, can cause muscular contractions, difficulty in breathing, reversible disturbances of heart function, but usually no organic damage.

Higher body currents inhibit internal muscle control, prejudicing the function of the muscles involved in the breathing process, thus causing asphyxia.

1.2.1 Ventricular Fibrillation

The ventricular fibrillation (V-fib) [11] is the loss of the normal heart rhythm. The V-fib causes the ventricles to *quiver*, or fibrillate, instead of contracting normally, preventing the heart from pumping blood and causing cardiac arrest. The ventricular fibrillation is the main cause of death in electric shock accidents.

The cardiac muscle, whose fibers have high contractile strength, specializes in pumping blood throughout the person's body. The contractions of the heart are stimulated by the *sinoatrial node* (SA), situated in the right atrium, which generates electrical impulses. The impulses propagate through the conductive tissues named *Bundle of His*, and *Purkinje fibers*, and reach the *atrioventricular node* (AV), situated in the center of the heart (Figure 1.1).

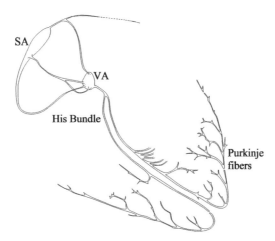

Figure 1.1 Electric conduction of the heart.

The Bundle of His, which departs from the AV, conducts the stimuli to the ventricles, which, after filling with blood, contract and push the blood through the arteries during the *systole*. After the contraction, the heart relaxes and fills up with blood again, awaiting further stimuli to contract again.

The net charge of the heart is zero; however, positive and negative charges are dynamically separated during each cardiac cycle and form an *electric dipole* vector that rotates and varies in magnitude with time. Thus, electric potential differences at different places along the person's body also change with time during each cardiac cycle, and this can be observed in an electrocardiogram (i.e., EKG or ECG) (Figure 1.2). Usually, *scalar* EKG measurements are performed: *vector* EKGs, which may more deeply describe the heart dipole rotation, are rarely performed. Typical potential differences showed in the EKG range between 30 and 500 μV.

During the P wave, the right and left atria contract; during the Q-R-S complex, the right and left ventricles contract (systole). The last event of the cycle, the S-T-U interval, is the repolarization of the ventricles: they return to the resting state; their walls relax and await the next signal. This complex procedure continues as the atria refill with blood and more electrical signals are sent by the SA; the heart-period duration is around 400 ms.

The superposition of external currents of larger magnitude to the normal bodily currents will override the control signals from the brain to the skeletal and cardiac muscles, which can no longer operate as intended, exposing persons to the risk of death.

The last half of the T wave is referred to as the *relative refractory period*, or the *vulnerable period*, which is known as the crucial time interval during which external electrical stimuli (i.e., electric shock) may induce the ventricular fibrillation.

For shock durations shorter than the cardiac cycle, the ventricular fibrillation may not occur, based on the lower probability that the external stimulus occurs during the vulnerable period of the heart.

For d.c. currents, experiments on animals and data derived from electrical accidents demonstrate that the threshold of V-fib for a downward current is

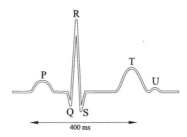

Figure 1.2 A normal electrocardiogram (EKG).

about twice as high as for an upward current; therefore, downward currents are less hazardous than upward currents.

For shock durations longer than the cardiac cycle (e.g., 0.5–1 s), the threshold of fibrillation for d.c. is several times higher than for a.c. However, for shock durations shorter than 200 ms, the threshold of d.c. fibrillation is approximately the same as for a.c. (measured in r.m.s. values).

1.2.2 The Heart-current Factor

The probability that the V-fib is induced is dependent upon the pathway of the body current. To compare the danger of different current paths through the body, standard IEC 60479-1[2] defines the *heart-current factor F* [12] (Eq. 1.1).

$$F = \frac{I_{ref}}{I_h} \tag{1.1}$$

I_{ref} is the body current that determines V-fib for the path left hand to feet, and I_h is the fibrillation current for different body paths, as shown in Table 1.1.

The larger is the heart-current factor; the more dangerous is the current pathway through the body.

As an example, a current of 225 mA hand-to-hand has the same likelihood of producing ventricular fibrillation as a current of 90 mA left hand-to-both feet. Therefore, the hand-to-hand pathway is less hazardous than the left hand-to-both feet.

Table 1.1 Heart-current factor *F* for different current paths

Path	Heart-Current Factor *F*	
	IEC	Simulated
Left hand–feet	1	1
Hands–feet	1	0.85
Left hand–right hand	0.4	0.75
Right hand–feet	0.8	0.88
Hands–seat	0.7	0.84
Left foot–right foot	0.04	0.01
Left hand–right foot	1.00	1.00
Left hand–left foot	1.00	1.00
Right hand–right foot	0.88	–
Right hand–left foot	0.8	0.89

---

2 (International Electrotechnical Commission) IEC/TS 60479-1: "*Effects of Current on Human Beings and Livestock - Part 1: General Aspects*".

The IEC does advise that the published heart-current factors must be considered as a rough estimation of the relative danger of the various current paths with regard to ventricular fibrillation. The IEC formulation of the heart-current factor is, in fact, based on experiments on corpses, animals and volunteers, or data from electrical accidents. Trials on animals produce results whose extrapolation to humans may not always be reliable, due to the obvious anatomical differences. In addition, exhaustive information about electrical accidents may not be available; therefore, it may not be possible to evaluate the magnitude of the body current affecting the injured and have reliable data.

A possible alternative computation for F can be obtained through *human phantoms*, which are computerized models that allow the numerical simulation [13] of the current pathways through the human body when subjected to external stimuli, as shown in the last column of Table 1.1.

The comparison between the two sets of heart-current factors shows that the IEC may underestimate the magnitude of F for pathways involving the right hand, probably due to the extrapolation of the results of measurements on animals to humans.

The same heart current factors are also applicable to d.c. currents.

1.3 The Electrical Impedance of the Human Body

The electrical impedance Z_B of the human body is capacitive in nature [14] due to the capacitance C_s of the skin (Figure 1.3); therefore, it depends on the frequency of the applied touch voltage. R_{Bi} is the internal body resistance; R_{cs} represents the resistance of the skin at the surface area of contact, which takes into account the

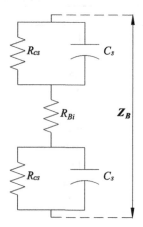

Figure 1.3 Impedances of the human body.

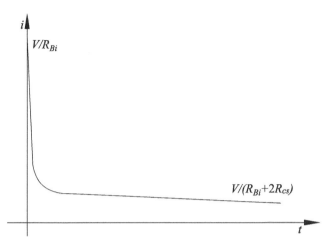

Figure 1.4 Current response of human body to d.c. voltage.

presence of the pores, which are small conductive elements; R_{cs} is strongly variable with environmental and physiological conditions (e.g., sweaty hands).

The model of Figure 1.3 has been validated on cadavers by analyzing the current response to a d.c. voltage V [15] (Figure 1.4).

When the d.c. touch voltage occurs, the capacitances C_s are not charged, and become short-circuits during the initial transient, bypassing the contact resistances R_{cs}; therefore, the ratio of the touch voltage V to the current peak equals R_{Bi}. After the transient expires, the capacitances of the skin become an open circuit, and the current reaches the steady-state value of $V / \left(R_{Bi} + 2R_{cs} \right)$, where the denominator is the total body resistance.

The impedance of the skin is the primary barrier against the flow of the body current, providing that the voltage is not high enough to puncture it (i.e., below 200 V), and the skin is not wet. Voltages greater than 200 V exceed the dielectric strength of the skin, C_s "fails" and short-circuits the contact resistance R_{cs}, reducing the resistance of the human body to R_{Bi}: the body current can cause greater damages to the internal organs.

In addition, at 50/60 Hz, the capacitive reactance of the skin is practically an open circuit, and $Z_B \approx R_{Bi} + 2R_{cs}$.

1.3.1 The Internal Resistance of the Human Body

R_{Bi} in Figure 1.3 represents the internal resistance of the human body, which depends on the chosen current pathway [16]. IEC 60479-1 expresses the resistance R_i of different segments of the body, ignoring the skin contribution at

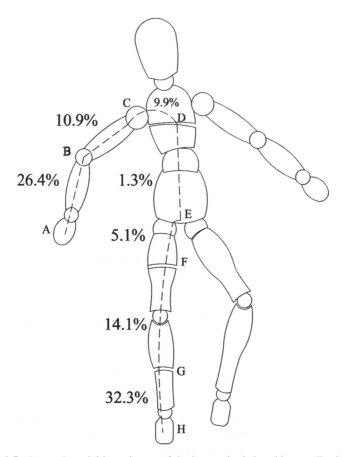

Figure 1.5 Internal partial impedances of the human body (no skin contribution).

50/60 Hz, as a percentage of the internal resistance of the human body related to the path hand-to-foot (Figure 1.5).

For example, the partial impedance of the trunk (i.e., segment D-E) is only 1.3% of the total body impedance hand-to-foot, due to large amount of conductive fluids normally present in the trunk.

The resistance R_i of the segment are determined as $R_i = \rho l / S$, where ρ is the segment tissue mean resistivity, l the mean length of the segment, and S its mean cross-sectional area. The cross-sectional area of the body segment plays a crucial role in determining its resistance: fingers and joints, such as elbow and knee, have higher resistance values due to their relatively small cross-sectional areas, even though they are made of well-conductive tissues.

The total body impedance for a given current path is obtained by adding the resistances R_i of the body segments for that path and the impedances of the skin at the surface areas of contact.

To underscore the role of the skin as the primary barrier against the flow of the body current, the US NIOSH[3] states that *"under dry conditions, the resistance offered by the human body may be as high as 100,000 Ω. Wet or broken skin may drop the body's resistance to 1,000 Ω"*.

IEC/TS 60479-1 affirms the variability of the body impedance Z_B and body resistance R_B related to the touch voltage, both a.c. (50/60 Hz) and d.c., and provides impedance values for the hand-to-hand pathway, in the case of dry skin and large contact areas (i.e., order of magnitude 100 cm^2), herein shown in Table 1.2.

Table 1.2 shows Z_B and R_B in the population percentile; for instance, for a touch voltage of 50 V, 95% of the population has an impedance of 4,600 Ω or less.

The body resistance for direct current (i.e., $f = 0$) is higher than the body impedance for alternating currents (i.e., $f = 50/60$ Hz) for touch voltages up to approximately 200 V, thanks to the blocking effect of the capacitances of the skin (i.e., they are open circuits at steady state); for a.c. contacts, the capacitances C_s are instead in parallel to the contact resistances R_{cs}.

For durations of current flow longer than 0.1 s, the skin will rupture, and Z_B approaches R_B.

The total body impedance Z_B depends on the area of contact with the energized part. Surface areas of contact are defined as *large*, *medium*, and *small*, with order of magnitude respectively of 100 cm^2, 10 cm^2, and 1 cm^2, and characterized by dry, water-wet, and saltwater-wet conditions.

Table 1.2 Body impedances and resistances for a current path hand-to-hand

Touch Voltage (V)	Z_B (Ω)			R_B (Ω)		
	5%	50%	95%	5%	50%	95%
25	1750	3250	6100	2100	3875	7275
50	1375	2500	4600	1600	2900	5325
150	850	1400	2350	875	1475	2475
200	800	1275	2050	800	1275	2050
225	775	1225	1900	775	1225	1900
400	700	950	1275	700	950	1275
500	625	850	1150	625	850	1150
1000	575	775	1050	575	775	1050

---

3 National Institute for Occupational Safety and Health. Publication 98-131: *"Worker Deaths by Electrocution - A Summary of NIOSH Surveillance and Investigative Findings"*.

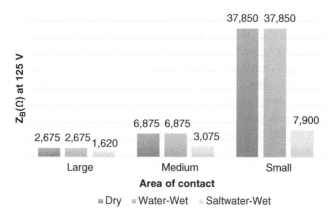

Figure 1.6 Body impedances at 1250 V for a path hand-to-hand vs. the area of contact.

In Figure 1.6, values of impedances not exceeded by the 95% of population, for a current path hand-to-hand and for a 125 V touch voltage (a.c. 50/60 Hz), are shown as a function of the surface areas of contact.

It can be observed that Z_B increases with polynomial law when the area of contact decreases. For a given area of contact, no appreciable differences in Z_B are present in dry and water-wet conditions for a touch voltage of 125 V.

The effect on Z_B of the surface area of contact increases when the touch voltage decreases; this is because touch voltages exceeding 200 V may rupture the capacitance of the skin and short-circuit the contact resistance.

In sum, Z_B is different from person to person and is dependent on several factors, including [17] but not limited to:

- the touch voltage;
- the supply frequency;
- the duration of the current flow;
- the conditions of wetness of the skin and surface area of contact;
- the general environment.

1.4 Thermal Shock

The current i passing through the human body during the contact time t with an energized part produces a physiological damage due to the generation and transfer of heat, per the Joule effect, to the biological tissues. An *electric burn* is defined as the burning of the skin or of an organ caused by the flow of an electric current along its surface or through it. Electric burn injuries account for 4% to 6% of all admissions to burn-care facilities [18].

The amount of heat generated in the tissue depends on the current density and the tissue conductivity. However, the conductivity is in turn determined by the heat generated in the tissue; since the tissue ionic conductivity increases with the increasing temperature, a further intensification in current density and temperature occurs. Thus, the thermal injury is determined as the result of a feedback mechanism; however, the conductivity change is not considered in burn models due to the complexity of the resulting nonlinear equations.

For the calculation of the amount of energy w delivered by a current i during the time t to a homogeneous volume of biological tissue of length l, cross-sectional area and ionic conductivity σ (Eq. 1.2), it is conservatively assumed an *adiabatic* process. Such a process calls for no heat removal into neighboring tissues by blood flow or by conduction and/or convection into the air, but it is presumed that all the heat stays within the tissue.

$$w = \frac{l}{\sigma S} i^2 t. \tag{1.2}$$

Vues of ionic conductivity of some biological tissues are listed in Table 1.3 [12].

The thermal injury will depend on the duration t of the contact and the temperature rise ΔT that the passing current will impose [19]. The thermal energy expressed in Eq. 1.2 is related to the temperature rise ΔT to which the volume of tissue is subjected. We can write the thermal balance equation (Eq. 1.3),

Table 1.3   Electrical conductivity σ of biological tissues

Tissue	σ (S/m)
Blood	7.00E-1
Bone	8.07E-2
Cartilage	1.71E-1
Fat	4.04E-2
Heart	8.27E-2
Kidney	8.92E-2
Muscle	2.33E-1
Nerve	2.74E-2
Skin (dry)	2.00E-4
Skin (wet)	4.27E-4

which describes the exchange of heat between the heat delivered w and the heat accumulated in the volume of tissue, assuming σ independent of the temperature.

$$w = \frac{1}{\sigma S} i^2 t = cS\Delta T \tag{1.3}$$

c is the volume-specific heat capacity, defined as the heat necessary to increase the temperature of a unit volume of a substance by 1°C.

The temperature rise ΔT is given by Eq. 1.4

$$\Delta T = \frac{J^2}{\sigma c} t \tag{1.4}$$

Equation 1.4 shows that the temperature rise ΔT depends on the square of the current density J and on the duration t of the current circulation.

The skin has the lowest conductivity among the biological tissues, and the current density is higher at the point of contacts on the body (referred to as *entry* and *exit* sites); therefore, for a given current, the highest temperature rise is achieved on the skin, which therefore suffers the greatest level of damage.

Thermal injury of the human skin occurs when the temperature rise persists for a sufficient length of time: for instance, to cause cutaneous injury, a temperature of the skin of 45°C requires a contact duration of 2 h; 51°C requires a contact duration of 2 min; and 60°C requires a contact duration of 3 s[4] Current densities of a few mA/mm^2 for a duration of about 0.5 to 1 s can cause burns, whereas a current density of a few tens of mA/mm^2 for a few seconds, will cause third-degree burns with destruction of deeper tissues and possible necrosis.

Thermal shock can also be caused by the heat released by electric arcs, which are accompanied by the vaporization of metal to form a superheated toxic gas.

1.5 Heated Surfaces of Electrical Equipment and Contact Burn Injuries

Burns can also be triggered by unintentional contact with hot surfaces of electrical equipment, which may be readily accessible during normal operations (e.g., the surface of a PV module) [20, 21].

---

4 Henriques, F.C. and Moritz, A.R. (Sep. 1947). Studies of Thermal Injury II. The Relative Importance of Time and Surface Temperature in the Causation of Cutaneous Burns. *The American Journal of Pathology* 23 (5): 695–720.

According to Stevens' Law, the perception of temperature is also *superlinear* with the stimulus, as varies as the 1.6 power of the temperature (i.e., experimentally analyzed by using a heated metal on a person's arm).

Most apparatus and appliances in industrial, commercial, and residential environments are thermally insulated unless the insulation would prevent their functions (e.g., the bottom surface of a flatiron). However, superficial temperatures of insulated equipment may still be high enough to cause burns from contact with readily accessible parts. The severity of such burns will depend on the thermal resistivity of the material of the touchable surface, and the pressure and duration of the contact.

An effective protection against burns can be established based on the acceptable contact period and on the level of acceptable injury.

According to the CENELEC Guide 29[5] for adults, a minimum contact period ranging between 0.5 s and 1 s should be used, based on the type of the equipment and where they are to be used (e.g., restrictive locations); however, extended reaction times may be considered based on the age of users that may possibly come into contact with the hot surface (Table 1.4).

Once the maximum operating temperature of readily accessible surfaces is determined, by direct measurement or calculation, the potential injury level may be established through the graph of Figure 1.7, which shows the relationship between surface temperature and exposure time.

The bottom curve T_B is the locus of the pairs temperature and exposure time representing the limit of the reversible epidermal injury; T_B describes the acceptable injury level as a first degree burn, that is, a burn where the temperature and/or duration are not sufficient to cause necrosis of the epidermis but

Table 1.4 Standard contact durations

Age Group	Exposure time (s)
Adults	0.5–1
age < 2 years	15
2 years < age < 6 years	4
6 years < age < 14 years	2
Elderly persons	1–4
Physical disabilities	According to nature of disability

---

5 NELEC Guide 29: "Temperatures of hot surfaces likely to be touched. Guidance document for Technical Committees and manufacturers". Ed. 1, 04-2007.

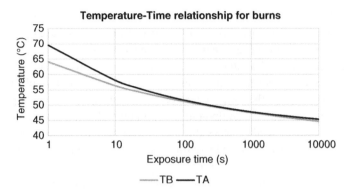

Figure 1.7 Temperature–Time Relationship for burns.

only reddening of the skin. T_A is the locus representing the complete trans-epidermal necrosis.

The surface of photovoltaic arrays in full sun can exceed the ambient temperature by 30°C or more, which may easily produce temperature greater than 60°C. It is therefore apparent that PV modules' surface with temperatures exceeding 64°C can only be contacted for 1 s before skin injury occurs.

IEC 60364-4-42[6] requires that accessible parts of electrical equipment within arm's reach do not attain temperatures exceeding the limits given in Table 1.8, to prevent burns caused by contact with heated surfaces.

The above temperature limits vary according to whether the part is intended to be hand-held or touched during normal use, and are based on the nature of the material of the accessible surface; they do not apply to equipment for which a maximum temperature is specified in the relevant product standard.

The temperature limits of Table 1.8 are rather large, and it would be prudent to be well below those values; if not possible, the equipment in question might be fitted with guards to prevent accidental contact.

1.6 Ground-Potential and Ground-Resistance

A ground electrode is a conductive part, embedded in the soil or in another conductive medium (e.g., concrete), which is in electrical contact with the earth [22].

---

6 IEC 60364-4-42: "*Low-voltage electrical installations – Part 4-42: Protection for safety – Protection against thermal effects*".

Table 1.8 Temperature limits in normal service for accessible parts of equipment

Accessible parts	Material of accessible surfaces	Maximum temperatures (°C)
A hand-held part	Metallic	55
	Non-metallic	65
A part intended to be touched but not hand-held	Metallic	70
	Non-metallic	80
A part that does not need to be touched for normal operation	Metallic	80
	Non-metallic	90

A connection to the ground can also be made through metalwork not forming part of the electrical installation, such as structural steelwork, metal, water supply pipes, or other buried metalwork. Such metalwork, however, should not be relied upon as an electrode, as it could be removed or replaced without any warning to users. The safety purpose of ground-electrodes is to effectively dissipate fault-currents into the soil.

To illustrate the relationship between ground-potentials, ground-resistances, and ground-currents, we study a hemispherical electrode, as this will allow the understanding of the performance of electrodes of different geometry.

We consider a hemisphere of radius r_0 embedded in a boundless and uniform soil of resistivity ρ, buried at a sufficient distance from a *receiving* electrode, and that the ground-current i leaking from this electrode flows radially into the soil (Figure 1.8).

The current density $\vec{J}$, identified as a vector quantity, through a surface S in the soil of infinitesimal thickness dl, placed at the distance r from the center of the hemisphere, is related to the uniform leakage current i through the flux operator expressed in Eq. 1.5.

$$i = \iint_S \vec{J} \cdot \vec{u}_n dS = J 2\pi r^2 \tag{1.5}$$

Equation 1.5 yields:

$$\vec{J} = \frac{i}{2\pi r^2}\hat{r}, \text{ for } r > r_o \tag{1.6}$$

where $\hat{r}$ is the unit vector in the radial direction.

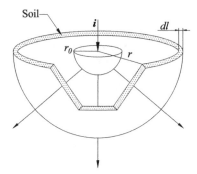

Figure 1.8 Hemispherical ground-electrode.

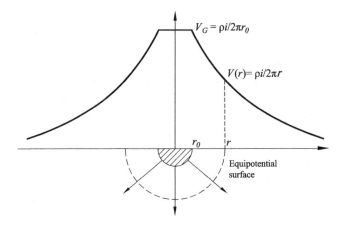

Figure 1.9 Hyperbolic distribution of the ground-potential $V(r)$ over the soil.

The electric field $\vec{E}$ at any distance r from the center of the hemisphere can be determined as:

$$\vec{E}(r) = \rho\vec{J}, \text{ for } r > r_o \tag{1.7}$$

The *ground-potential* on the soil surface at any distance r from the center of the hemisphere, which is taken zero at infinity, is:

$$V(r) = -\int_{\infty}^{r}\vec{E}\cdot d\vec{r} = -\rho\int_{\infty}^{r}\vec{J}\cdot d\vec{r} = -\rho\frac{i}{2\pi} = \int_{\infty}^{r}\frac{1}{r^2}\cdot d\vec{r} = \rho\frac{i}{2\pi r} \tag{1.8}$$

The ground-potential $V(r)$ features a hyperbolic distribution through the soil, with the coordinate axes as asymptotes (Figure 1.9).

The equipotential surfaces are hemispheres, including the actual surface of the electrode. Points belonging to the same equipotential surface have equal potential both on the surface and deep in the soil. Current lines are perpendicular to such surfaces.

The *ground-potential rise* on the surface of the hemisphere, that is, the potential at the distance r_0 from its center, is

$$V_G = V(r_0) = -\int_{\infty}^{r_0} \vec{E} \cdot d\vec{r} = -\rho \int_{\infty}^{r_0} \vec{J} \cdot d\vec{r} = -\rho \frac{i}{2\pi} = \int_{\infty}^{r_0} \frac{1}{r^2} \cdot d\vec{r} = \rho \frac{i}{2\pi r_0} \quad (1.9)$$

We define the resistance R_G of the hemisphere-electrode to earth (from now on the *ground-resistance*) as the ratio of the ground-potential rise V_G to the leakage current i (Eq. 1.10).

$$R_G = \frac{\rho}{2\pi r_0} \quad (1.10)$$

The ground-resistance of a ground-electrode can be seen as an equivalent one-port (Figure 1.10): one terminal of the one-port is the metal connection to the electrode (generally named *grounding electrode conductor* in codes and standards), whereas the other terminal represents a point at zero potential (i.e., a point at sufficient distance from the electrode where the potential is negligible).

The ground symbol (from IEC 60417 "*Graphical symbols for use on equipment*," symbol 5017) does not represent the soil, but a point at sufficient distance from the electrode where the surface potential is negligible.

From the graph of the ground potential of Figure 1.9, it can be observed that the radius r_0 of the hemisphere identifies the point from where the hyperbolic distribution starts. For a given hemisphere, different values of the product ρi determine different hyperbolae, whose distance for the horizontal axis depends on the soil resistivity and the fault-current.

The rate-of-change of the potential with the distance r from the hemisphere (i.e., the potential gradient) is defined in Eq. 1.11.

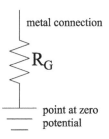

metal connection

R_G

point at zero potential

Figure 1.10 Electrode ground-resistance as an equivalent one-port.

$$\frac{dV}{dr} = -\frac{\rho i}{2\pi r^2}$$ (1.11)

which shows that the maximum variation of the ground potential occurs in proximity of the hemispherical electrode (i.e. $r \approx r_0$).

1.6.1 Area of Influence of a Ground-electrode

The electric field is a *long-range* field and is zero only at infinite distance from its source, and so is the *ground potential*. In engineering practice, however, the design of ground electrodes is based on the *area of influence*, which defines the zone beyond which the ground potential can be considered negligible. If we evaluate the ground potential at the distance $r = 5r_0$, we obtain:

$$V(5r_0) = \frac{V_G}{5}$$ (1.12)

At a distance $5r_0$ from the center of the hemisphere, the ground potential reduces to 20% of the ground potential rise, and this result has a general validity, independently of the shape of the electrode. It can conventionally be assumed that the hemispherical volume of the earth of radius $5r_0$ is the area of influence of the electrode. For differently-shaped electrodes (e.g., rods, rings, grids, etc.), their maximum dimensions can be used in lieu of the radius; for instance, for grounding grids, the largest diagonal can be employed to identify the area of influence.

Two unconnected ground-electrodes are defined as independent from each other if they are outside of their respective areas of influence.

1.7 Hemispherical Electrodes in Parallel

Two identical hemispherical electrodes of radius r_0 are connected in parallel into a uniform soil of resistivity ρ, and each leaks the current $i/2$. The electrodes are buried at a distance d.

The hemispheres attain the same ground-potential rise, which can be calculated by superimposing the potentials of each electrode, supposed isolated from the other (Eq. 1.13).

$$V_G = \rho \frac{i/2}{2\pi r_0} + \rho \frac{i/2}{2\pi(d - r_0)} = \rho \frac{i}{4\pi r_0} \frac{1}{1 - \dfrac{r_0}{d}}$$ (1.13)

Thus, the total ground resistance is given by Eq. 1.14.

$$R_G = \frac{\varrho}{4\pi r_0} \frac{1}{1 - \dfrac{r_0}{d}} \tag{1.14}$$

If $d > 5r_0$, the second factor in Eq. 1.14 is ≈ 1, and R_G is mathematically expressed by the formula of the parallel of the ground resistances of each electrode.

If the electrodes are closer than $5r_0$, they are not independent from each other, and their total resistance is greater than their mere parallel (Figure 1.11).

1.8 Hemispherical Electrodes in Series

Let us consider two identical hemispheres embedded into the soil at the distance $d \geq 5r_0$ center-to-center, so that they can be considered independent from each other. The first electrode leaks the current i, whereas, the second electrode receives the same current (Figure 1.12).

The resulting surface potential $V(r)$ at a generic point r along a line joining the two hemispheres is given by the superposition of the potentials imposed by each electrode (Eq. 1.15).

$$V(r) = \rho \frac{i}{2\pi r} - \rho \frac{i}{2\pi (d-r)} = \rho \frac{i}{2\pi} \left(\frac{1}{r} - \frac{1}{d-r} \right). \tag{1.15}$$

If $r = d/2$, which is the center point between the two hemispheres, the value of the surface potential is $V(d/2) = 0$.

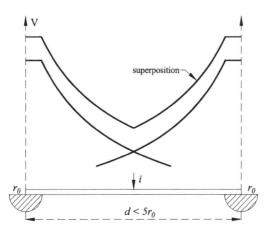

Figure 1.11 Ground-electrodes in parallel.

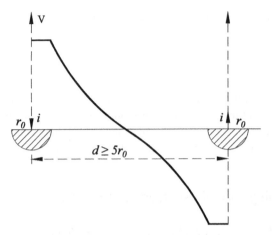

Figure 1.12 Hemispherical electrodes connected in series.

The point of the soil at which the ground potential is zero is the location that must be identified for the correct measurement of the ground resistance R_B of an electrode, as further elaborated.

1.9 Person's Body Resistance-to-ground and Touch Voltages

In the case of the pathway hands-to-feet, the current will flow into the soil through the feet, and its amount will depend on the series between the body resistance R_B and the person's resistance-to-ground R_{BG}. The person's resistance-to-ground limits the circulation of the body current, therefore is beneficial for the electrical safety of individuals.

For an approximate calculation of R_{BG}, the adult human foot can be modeled as a circular plate of radius $r_f = 0.08$ m, laying on a surface of resistivity ρ. The expression in ohms of the ground resistance R_f of such electrode is given in Eq. 1.16.

$$R_f = \frac{\rho}{4r_f} \cong 3\rho \tag{1.16}$$

If we assume that the two feet act as ground electrode in parallel, and that the plates do not interfere with each other, the body resistance-to-ground R_{BG} equals 1.5ρ.

In general, it may be conservatively assumed that the person does not wear shoes or gloves, and that there is no floor to limit the body current. However, standard EN 50522[7] allows, for calculation purposes, additional known resistances in series to the body resistance (e.g., gloves, footwear, standing surface made of insulating material, such as gravel, asphalt, etc.); EN 50522 identifies in 1 kΩ the average value for old and wet shoes. Typical shoe resistances are 5–10 kΩ for wet leather soles, 100–500 kΩ for dry leather soles, and 20 MΩ for rubber soles.

To calculate the body current, we determine the parameters of a Thevenin equivalent circuit, V_{th} (i.e., equivalent voltage source) and R_{th} (i.e., equivalent resistance), as seen from the point of contact of the body with the energized part, and the ground (Figure 1.13).

R_{th} is generally negligible if compared to $R_B + R_{BG}$, and therefore can be conservatively ignored; the fault source can be thought of as an ideal voltage generator.

In these conditions, the body current i_B can be calculated with Eq. 1.17.

$$i_B = \frac{V_{th}}{R_B + R_{BG}} \qquad (1.17)$$

In Figure 1.12, V_{ST}. is the *prospective touch voltage*, which is defined as the potential difference between simultaneously accessible conductive parts, when those conductive parts are not being touched by a person. V_T is the (effective) *touch voltage*, defined as the potential difference between accessible conductive parts when touched simultaneously by a person (one part can be the ground) [23–27]. The value of the effective touch voltage is affected by the persons' body resistance, greatly variable, and the person's resistance-to-ground. Consequently, the same touch voltage may correspond to different body currents, and this makes the touch voltage a rather ineffective indicator of hazard.

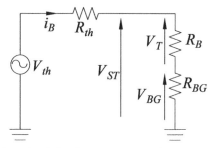

Figure 1.13 Equivalent circuit for the computation of body currents due to a touch voltage.

---

7 EN 50522:2010-11: *"Earthing of power installations exceeding 1 kV a.c"*.

Thus, for the purpose of designing protective measures against electric shock, technical standards identify a conventional body resistance of 1 kΩ.

The *step voltage* is defined as the voltage between two points on the earth's surface that are 1 m distant from each other, which is considered the standard stride length of a person.

In the worst-case scenario, prospective touch voltages may equal the ground potential rise V_G. To better clarify the concept, let us assume that in the event of a fault, a hemisphere of radius r_0, and resistance-to-ground R_G leaks to ground the current i. Let us also assume a person standing in a region at zero potential; the person is touching a metallic structure electrically connected to the hemisphere for grounding purposes (Figure 1.14).

The hemisphere attains the ground potential rise $V_G = iR_G$, and so does the metallic structure: the person's hand is at the potential V_G, whereas their feet are at zero potential. The prospective touch voltage V_{ST} equals the ground potential rise; however, the effective touch potential V_T is a lower value, equal to the voltage drop on the person's body resistance R_B, as established by the voltage divider between R_B and R_{BG} (Figure 1.13).

The distribution of the ground potential $V(r)$ over the soil is shown in Figure 1.15. It can be seen that in correspondence with the persons' feet, the surface ground-potential rises up from (almost) zero to the value V_{BG}, which accordingly lowers the touch voltage.

The other possible scenario is the person standing in a non-zero potential region, at a distance r from the center of the hemisphere less than $5r_0$ (Figure 1.16).

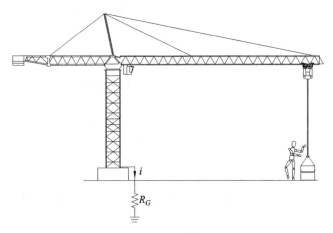

Figure 1.14 Person standing in a region at zero potential.

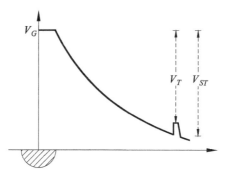

Figure 1.15 Distribution of the ground-potential with a person standing in a region at zero potential.

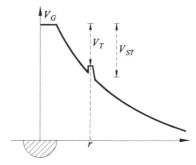

Figure 1.16 Distribution of the ground-potential with person standing in a region at non-zero potential.

In this case, the prospective touch voltage V_{ST} is less than V_G, even though the person's hand is still at the potential V_G. The person's feet, in fact, will be at a higher potential, with an evident reduction of both prospective and touch voltage.

A possible hazardous situation is when the person, although standing in a non-zero potential area, may also be in contact with a conductive part at zero potential. Such parts, defined as *Extraneous-Conductive-Parts* (EXCPs) [28, 29], may include water pipes, exposed metallic structural parts of the building, etc.

EXCPs may have a very low resistance-to-ground R_{EX}, such that $R_{EX} \ll R_{BG}$. therefore, if contacted, they may completely bypass the person's body resistance-to-ground. The person is therefore subjected to a greater touch voltage, and if R_{EX} were zero, the touch voltage would equal V_{ST} (Figure 1.17).

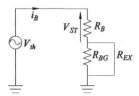

Figure 1.17 Equivalent circuit for the computation of body currents in the presence of an EXCP.

Example 1.1 A hemispherical electrode of radius $r_0 = 2$ m is buried in the ground of resistivity $\rho = 200$ Ωm and connected to a tower crane (Figure 1.13). A person is standing 10 m away from the center of the hemisphere and holding a metal hook electrically connected to the hemisphere. A ground-fault causes a current $i = 100$ A to flow into the earth through the hemisphere. Determine the electric current i_B through the person's body in the case of touch, assuming a conventional body resistance of 1 kΩ.

Solution

From Eq. 1.9, the ground potential rise of the hemisphere is:

$$V(r_0) = \rho \frac{i}{2\pi r_0} = 1{,}592.4 \text{ V}$$

The touch voltage is $V(r_0) - V(r)$, where $V(r)$ can be calculated with Eq. 1.8 with $r = 10$ m.

$$V(r) = \rho \frac{i}{2\pi r} = 477.7 \text{ V}$$

$$V_t = V(r_0) - V(r) = 1114.6 \text{ V}$$

From Eq. 1.16, the body current with $R_B = 1$ kΩ, and $R_{BG} = 1.5\rho = 300$ Ω, is:

$$iB = \frac{v_{th}}{R_B + R_{BG}} = 857.4 \text{ mA}$$

1.10 Identification of *Extraneous-Conductive-Parts*

With the purpose of reducing touch voltages, applicable standards require that in each installation main protective bonding conductors connect to the main ground busbar the EXCPs, which may include: metal installation pipes for gas,

water, heating, etc.; non-insulating floors and walls; exposed metallic structural parts of the building. These items are mere examples of conductive parts
that may require the protective equipotential bonding; the identification of
EXCPs is crucial for safety and a verification if conductive parts are EXCPs is
necessary before connecting them to the ground busbar.

BS 7671[8] defines an EXCP as "*a conductive part liable to introduce a potential,
generally earth potential, and not forming part of the electrical installation.*"

The term *earth potential* is generally assumed to be zero volts and introduced
by the general mass of the earth into the installation. This definition implies
that such conductive parts would originate outside the building and be in
contact with the ground. If this can be verified by visual inspection, the part in
question should be bonded as close as practicable to their point of entry within
the building. It is important to clarify that for bonding purposes, the point of
connection of pipework should take place along the section of the pipe from
the meter into the building, which is owned by the user, and not on the section
that comes in from the road into the meter. This prevents corrosion issues to
the service pipework.

If it is not possible to verify by visual inspection alone that a conductive part
is an EXCP, a measurement of the resistance R_{EX} between the conductive part
in question and the main ground busbar should be performed. If the measured
resistance R_{EX} satisfies Eq. 1.18, based on the circuit of Figure 1.16, the conductive part in question is not to be considered an EXCP.

$$R_{EX} \geq \frac{V_{th}}{I_{BM}} - R_B \tag{1.18}$$

V_{th} is the nominal voltage to ground of the installation (in V); R_B is the standard value of body resistance of 1 kΩ; I_{BM} (A) is the maximum value of body
current that is deemed acceptable to the designer. Threshold values for I_{BM}
may be 0.5 mA (i.e., the threshold of perception), 10 mA (the threshold of
let-go), or 30 mA.

However, the designer should assess if the measured resistance to the ground
of the concerned conductive part may change (i.e., decrease) throughout the
lifetime of the installation.

If, for example, the chosen threshold of safe current is 30 mA with
$V_{th} = 230$ V, and the measured resistance between a conductive part not
forming part of the electrical installation (a presumed EXCP) and the ground is
greater than 6.67 kΩ, then no connection to the main ground terminal of the
metal part in question would be required.

---

8 BS 7671:2018 "Requirements for Electrical Installations".

It is important to recognize that the indiscriminate connection to the ground terminal of mere conductive parts that are not EXCPs may cause fault potentials to be transferred throughout the installation. This can cause the risk of electric shock for persons standing outside of the installation that is in contact with the presumed EXCP that has become live (e.g., a metallic fence).

1.11 Measuring Touch Voltages

Measurements after construction of the installation can verify the adequacy of the electrical design of the substation [30]. Measuring touch voltages [31–33] includes two choices: measure the prospective touch and step voltages, by using a high-impedance voltmeter, or measure the effective touch and step voltages occurring across an appropriate resistance of 1 kΩ, which represents the human body.

For touch voltage measurements a current injection method may be used (Figure 1.18).

An alternating voltage of approximately the system frequency is applied between the facility ground electrode and an auxiliary ground electrode, located far enough to guarantee separate zones of influence (e.g., 4 or 5 times the maximum dimension of the facility ground electrode). A test current i_m is injected into the facility grounding system, which causes a measurable ground potential rise.

The test current should be so high that the measured touch voltage, referred to as the test current, is greater than possible disturbance voltages; according to EN 50522, this may be ensured for test currents of at least 50 A. Always according to EN 50522, the measuring electrodes for the simulation of the feet, connected in parallel, must have a total area of 400 cm², lie on the ground with a minimum

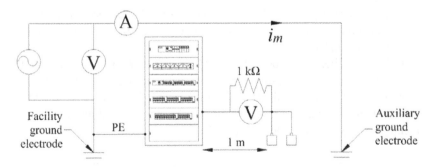

Figure 1.18 Touch voltage measurement with the current injection method.

total force of 500 N, and placed at a distance of 1 m from the equipment of the installation, or from an EXCP. During the test, the auxiliary ground electrode may assume a dangerous ground potential rise, and therefore should be guarded.

The tip-electrode for the simulation of the hand touching the equipment, or an EXCP, must be capable of penetrating a paint coating (not insulation).

The reading of the voltmeter, which is referred to as the test current, must then be scaled up by multiplying it by the ratio of the effective ground-fault current provided by the utility to the test current.

The touch voltage measurement should be performed in the substation, as a sampling test, keeping in mind that higher magnitudes for the touch voltages may be found around the edge of the ground grid.

References

1 Moreno, B., and López, A.J. (2008). The effect of renewable energy on employment. The case of Asturias (Spain). *Renewable and Sustainable Energy Reviews* 12 (3): 732–751.

2 Bulavskaya, T., and Reynès, F. (2018). Job creation and economic impact of renewable energy in the Netherlands. *Renewable Energy* 119: 528–538. doi:9.

3 Sooriyaarachchi, T.M., Tsai, I.-T., El Khatib, S., Farid, A.M., and Mezher, T. (2015). Job creation potentials and skill requirements in, PV, CSP, wind, water-to-energy and energy efficiency value chains. *Renewable and Sustainable Energy Reviews* 52: 653–668.

4 Gammon, T., Lee, W., and Intwari, I. (2019). Reframing our view of workplace 'electrical' injuries. *IEEE Transactions on Industry Applications* 55 (4): 4370–4376.

5 Crow, D.R., Liggett, D.P., and Scott, M.A. (2017). Changing the electrical safety culture. *2017 IEEE IAS Electrical Safety Workshop (ESW)*, 1–7.

6 Neitzel, D.K. (2016). Electrical safety by design and maintenance. *2016 IEEE Pulp, Paper Forest Industries Conference (PPFIC)*, 6–13.

7 Mathe, L., Sera, D., Spataru, S.V., Kopacz, C., Blaabjerg, F., and Kerekes, T. (2015). Firefighter safety for PV systems: A solution for the protection of emergency responders from hazardous dc voltage. *IEEE Industry Applications Magazine* 21 (3): 75–84.

8 Gordon, L.B., Cartelli, L., and Graham, N. (November 2018). A complete electrical shock hazard classification system and its application. *IEEE Transactions on Industry Applications* 54 (6): 6554–6565.

9 John Cadick, P.E., Mary Capelli-Schellpfeffer M.P.A., M.D., Neitzel, C.P.E.D.K., and Winfield, A. (2012). *Electrical Safety Handbook*, 4th ed. New York: McGraw-Hill Education.

10 Winfield, A., Capelli-Schellpfeffer, M., Neitzel, D., and Cadick, J. (2012). *Electrical Safety Handbook*, 4th ed. McGraw-Hill Professional.

11 Rachford, J. (2020). Explaining ventricular fibrillation in simple electrical terminology. *2020 IEEE IAS Electrical Safety Workshop (ESW)*, 1–3.

12 Freschi, F., Guerrisi, A., Tartaglia, M., and Mitolo, M. (2013). Numerical simulation of heart-current factors and electrical models of the human body. *IEEE Transactions on Industry Applications* 49 (5): 2290–2299.

13 Freschi, F., and Mitolo, M. (2017). Currents passing through the human body: The numerical viewpoint. *IEEE Transactions on Industry Applications* 53 (2): 826–832.

14 De Santis, V., Beeckman, P.A., Lampasi, D.A., and Feliziani, M. (February 2011). Assessment of human body impedance for safety requirements against contact currents for frequencies up to 110 MHz. *IEEE Transactions on Biomedical Engineering* 58 (2): 390–396.

15 Jiang, H., and Brazis, P.W. (2018). Experiments of DC human body resistance I: Equipment, setup, and contact materials. *2018 IEEE Symposium on Product Compliance Engineering (ISPCE)*, 1–6.

16 Boron, S., Heyduk, A., Joostberens, J., and Pielot, J. (2016). Empirical model of a human body resistance at a hand-to-hand DC flow. *Elektron. Ir Elektrotechnika* 22 (4): 26–31.

17 Lavrova, O., Quiroz, J.E., Flicker, J., and Gooding, R. (2017). Updated evaluation of shock hazards to firefighters working in proximity of PV systems. *2017 IEEE 44th Photovoltaic Specialist Conference (PVSC)*, 1280–1285.

18 Alam, M.K., Khan, F., Johnson, J., and Flicker, J. (2015). A comprehensive review of catastrophic faults in PV arrays: Types, detection, and mitigation techniques. *IEEE Journal of Photovoltaics* 5 (3): 982–997.

19 Qiu, L., and Wang, C. (2011). Research on the relationship between skin burn and action time. *Proceedings 2011 International Conference on Human Health and Biomedical Engineering*, 1201–1203.

20 Yen, M., Colella, F., Kytomaa, H., Allin, B., and Ockfen, A. (2020). Contact burn injuries: Part I: The influence of object thermal mass. *2020 IEEE Symposium on Product Compliance Engineering - (SPCE Portland)*, 1–5.

21 Yen, M., Colella, F., Kytomaa, H., Allin, B., and Ockfen, A. (2020). Contact burn injuries: Part II: The influence of object shape, size, contact resistance, and applied heat flux. *2020 IEEE Symposium on Product Compliance Engineering - (SPCE Portland)*, 1–6.

22 He, J., Zeng, R., and Zhang, B. (2013). Fundamental concepts of grounding. In: *Methodology and Technology for Power System Grounding*, 1–26. IEEE, Wiley.

23 Mitolo, M. and Liu, H. (2016). Touch voltage analysis in low-voltage power systems studies. *IEEE Transactions on Industry Applications* 52 (1): 556–559.

24 Laukamp, H. and Bopp, G. (July 1996). Residential PV systems—electrical safety issues and installation guidelines. *Progress in Photovoltaics: Research and Applications* 4 (4): 307–314.

25 Hernández, J.C. and Vidal, P.G. (2009). Guidelines for protection against electric shock in PV generators. *IEEE Transactions on Energy Conversion* 24 (1): 274–282.

26 Kontargyri, V.T., Gonos, I.F., and Stathopulos, I.A. (November 2015). Study on wind farm grounding system. *IEEE Transactions on Industry Applications* 51 (6): 4969–4977.

27 Karegar, H.K. and Arabi, M. (2010). New wind turbine grounding system to reduce step touch voltage. *2010 IEEE International Conference on Power and Energy*, 168–171.

28 Colella, P., Pons, E., and Tommasini, R. (2017). Dangerous touch voltages in buildings: The impact of extraneous conductive parts in risk mitigation. *Electric Power Systems Research* 147: 263–271.

29 Mauromicale, G., Raciti, A., Rizzo, S.A., Susinni, G., Parise, G., and Parise, L. (2019). E-mobility: Safety, service continuity and penetration of charging systems. *2019 AEIT International Conference of Electrical and Electronic Technologies for Automotive (AEIT AUTOMOTIVE)*, 1–6.

30 Mitolo, M. and Bajzek, T.J. (2018). Measuring the electrical safety in low-voltage distribution systems. *2018 IEEE/IAS 54th Industrial and Commercial Power Systems Technical Conference (I CPS)*, 1–4.

31 Kosztaluk, R., Dinkar Mukhedka, R., and Gervais, Y. (November 1984). Field measurements of touch and step voltages. *IEEE Transactions on Power Systems* PAS-103 (11): 3286–3294.

32 Meliopoulos, A.P.S., Patel, S., and Cokkinides, G.J. (1994). A new method and instrument for touch and step voltage measurements. *IEEE Transactions on Power Delivery* 9 (4): 1850–1860.

33 Wohlgemuth, J.H. and Kurtz, S.R. (2012). How can we make PV modules safer?. *2012 38th IEEE Photovoltaic Specialists Conference*, 3162–3165.

2

Safety-by-Design Approach in AC/DC Systems

The most beautiful sea hasn't been crossed yet.
The most beautiful child hasn't grown up yet.
The most beautiful days we haven't seen yet.
And the most beautiful words I wanted to tell you
I haven't said yet.

Nazim Hikmet

2.1 Introduction

A photovoltaic (PV) system consists [1, 2] of PV generators (i.e., arrays of panels) and a power conversion equipment (i.e., the inverter). A PV *module* is the smallest complete environmentally protected assembly of interconnected cells; PV modules

Electrical Safety Engineering of Renewable Energy Systems. First Edition. Rodolfo Araneo and Massimo Mitolo.
© 2022 by The Institute of Electrical and Electronics Engineers, Inc. Published 2022 by John Wiley & Sons, Inc.

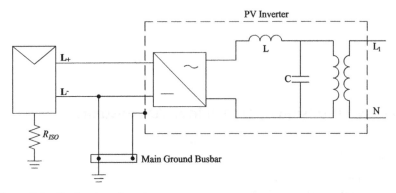

Figure 2.1 Single-phase inverter.

can be connected in series to form a *string*; a PV *array* may consist of a single *module*, a single *string*, or several parallel-connected strings.

The inverter [3–5] converts the output d.c. voltage of the PV generators, into an a.c. waveform that can be applied to the power grid, and allows the PV system to function at its maximum power point. Inverters may use the *Pulse Width Modulation* (PWM), which allows changing the magnitude of the a.c. voltage and its frequency, if required. The converted output voltage waveform may not be perfectly sinusoidal [6], due to the presence of harmonics [7, 8]. Thus, a transformer and an LC filter, placed between the conversion stage and the power grid, are normally employed to suppress the undesired harmonic content, as well as to match the power grid voltage (Figure 2.1).

The transformer provides galvanic isolation between the power grid and the PV systems, thus increasing the electrical safety of persons, and also prevents direct currents from being injected into the power grid, which could damage equipment and saturate the cores of the distributor's distribution transformers. However, the presence of the transformer [9, 10], especially the low-frequency type, reduces the efficiency of the PV generator, due to its core losses.

It is apparent that the PV equipment on the d.c. side of the inverter is always energized, even when the system is disconnected from the power grid.

The protection against ground-faults occurring in PV arrays is based on the following fundamental PV system topologies:

- the PV array may or may not be electrically separated from the grounded a.c. system;
- the PV array may or may not be functionally grounded;
- the AC system to which the PV inverter output connects may or may not be grounded.

The protection against electric shock is based on the system grounding, as well as on an effective equipotential bonding.

Two types of system grounding may be employed: grounded PV systems, generally adopted in the United States per the National Electrical Code (NEC, section 690), which requires that for a PV source with a system voltage greater than 50 V, one conductor of the 2-wire system (i.e., negative or positive) be grounded; ungrounded PV systems, which requires the *live* parts to be ungrounded and that the equipment (e.g., the PV inverter chassis) be grounded, per international standards (i.e., IEC).

Regardless of the type of system grounding, a good insulation resistance R_{iso} between live parts of the PV modules and the ground is of crucial importance to prevent leakage currents and reduce electric shock hazards. The insulation resistance [11, 12] may greatly decay in wet operating conditions, and/or due to the aging of the insulation compound; if it falls below a certain threshold (e.g., 20 kΩ), insulation monitors may shutdown the inverters, thereby taking offline portions of the PV generators. It is safe to say that R_{iso} is inversely proportional to the PV system rated power: modules, cables, and connectors have an inherent insulation resistance to ground, which appears in parallel with the others. Thus, R_{iso} goes down as the PV system power goes up.

2.2 Class I PV Equipment

PV equipment (i.e., PV generators, inverters, etc.) may be rated as *Class I*, that is, an apparatus with the basic insulation as the provision for the basic protection, and the protective bonding as the provision for the fault protection.

Metal frames of Class I PV arrays must be bonded to the main ground bus; the PV array mounting racks may be extraneous-conductive-parts, if their resistance-to-ground falls below a certain value, as discussed in Chapter 1. In this

a)   b)

Figure 2.2 (a) Metal frames and mounting racks of PV arrays and (b) Equipotential bonding connection.

Figure 2.3 Sun-tracking PV arrays.

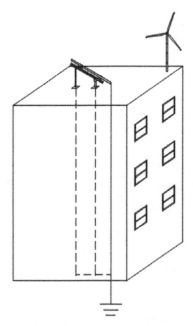

Figure 2.4 Mounting racks are equipotential to metal frames of roof-mounted PV generators.

case, PV modules must be in good contact with the metal frames so that to reduce touch voltages in the case of failure of their basic insulation (Figures 2.2 and 2.3).

Sun-tracking PV arrays follow the path of the sun, increasing the energy collection. Active tracking mounts employ electric motors linked to light sensors that automatically move the array toward the sun. This equipment is normally Class I, and the protection against electric shock is achieved thanks to the partnership between protective devices and protective conductor to promptly disconnect the supply in the event of faults.

In general, the mounting racks of rooftop mounted PV generators do not require additional equipotential bonding connections to the structure of the building itself. The racks are, in fact, in good electrical contact with the building, which is equipotential to the intentional grounding system connecting the metal frame of the PV modules (Figure 2.4).

In the case of failure of the insulation of the PV panels, racks and frames are at the same potential, regardless of the value of the actual resistance of the building structure, through which no-fault current will flow.

2.3   Class II PV Equipment

Alternatively, PV systems may be rated as *Class II* equipment (cables included) with *basic* insulation as provision for basic protection, and *double*, or *reinforced* insulation as provision for fault protection.[1] Touchable conductive parts of Class II PV equipment must not be intentionally connected to a protective conductor. The probability that a single failure can result in dangerous voltage is considered very low when compared to the probability that the enclosure might become live due to transferred voltages conveyed by the protective conductor. However, Class II PV equipment may be provided with means for the connection to ground for *functional* (as distinct from *protective*) purposes (e.g., installation of *Insulation Monitoring Devices*). Thus, metal frames and mounting racks of PV arrays may be outfitted with terminals insulated from live parts by double or reinforced insulation for the functional ground.

According to BS 7671,[2] protection by Class II, or equivalent insulation, must be preferably adopted on the d.c. side.

2.4   Ground Faults and Ground Fault Protection

According to IEC *International Electrotechnical Vocabulary* (IEV ref. 614-02-16), a ground fault consists of an *accidental conductive path between a live conductor and the ground*. Such path can pass through a faulty insulation, but also

---

1 IEC 61140: "Protection against electric shock - Common aspects for installation and equipment".

2 BS 7671: "*Requirements for Electrical Installations. IET Wiring Regulations*".

through structures or through vegetation. A ground fault with negligible fault impedance may be named as a *short circuit to ground*.

The major functions of the PV ground fault protection system (PV-GFP) [13, 14] against shock hazard include:

- The detection of the first ground fault in grounded, or functionally grounded PV systems, and the removal of the ground reference, which is a return path for the touch current within the safe time, in accordance with the conventional time-current curves describing the effects of d.c. current on persons. The permissible magnitude of the touch current is based on the physiological effects of the d.c. current passing through the human body and are conventionally set to values ranging between 30 mA to 90 mA.
- The detection of the first ground fault in non-ground-referenced (or *ungrounded*) PV systems, within 24 h. As later discussed, the first fault in ungrounded systems produces a low magnitude fault current and is not immediately hazardous. However, the detection of the first fault in a reasonable time is vital to reduce the probability that the occurrence of *second* fault could cause the main array d.c. switch to open and disconnect the d.c. circuit to the inverter, preventing possible damages to its electronics, as possibly required by inverter's product standards.

It is important to note that the array insulation resistance to ground R_{iso} may offer another path to the ground current. When R_{iso} decreases (e.g., due to increase in the PV system size), the disconnection of the intentional ground reference by the PV-GFP may no longer be an effective protection against electric shock. The lowered R_{iso}, in fact, may still allow the circulation of a ground current equal to, or in excess of, the permissible values, even though the system has been ungrounded by the PV-GFP.

The PV modules are "uncontrollable" sources at the origin, because if there is light, they cannot be deenergized: protective devices downstream the PV array always leave the system energized.

In addition, PV devices produce fault currents almost indistinguishable from normal operating current, which render overcurrent devices ineffective as a protection against abnormal currents in the *forward* direction. Protective devices (i.e., fuses and circuit breakers) are, therefore, installed at the source of the overcurrent, which is where the PV string source circuits are combined (i.e., the combiner box). String protective devices are sized according to the maximus current the PV module can withstand (e.g., 15 A), and they detect the abnormal current from the parallel-connected PV sources *backfeeding* the fault (Figure 2.5).

If a PV string experiences a ground fault, the remaining $n - 1$ strings, paralleled through combiner fuses, will backfeed the faulted PV source circuit, the faulted string's combiner fuse will see currents flowing in the reverse direction

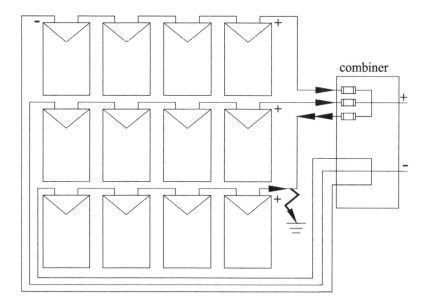

Figure 2.5 Protective devices in combiner box.

and will open and isolate the string. However, when the fuse opens, the ground fault current's magnitude is lowered to the single-string contribution, which may be below the setting of PV-GFP. Therefore, if the operation of the fuse precedes the operation of the PV-GFP, the ground fault may become a *latent* fault, which can go undetected and exist in the system even for months without repair. These undetected faults may not produce an immediate danger, being the fault current low enough, but a second fault becomes increasingly likely; the second fault could increase the fault current to hazard levels and is a scenario that must be avoided by employing improved PV-GFP strategies.

PV arrays are by their nature located outdoors; therefore, environmental conditions should be taking into consideration in selecting PV-GFP settings to avoid possible nuisance tripping, extensive recovery times, and loss of revenue. Environmental factors may affect the fault-free resistance to ground R_{iso} of the PV array, which may greatly vary between cold, damp overnight conditions and warm daytime conditions, and dry and wet weather.

2.5   Functionally Grounded PV Systems

The functional grounded system (Figure 2.6) has one conductor of the PV array intentionally connected to ground for purposes other than safety, and by means not necessarily sized in compliance with safety requirements. The functional

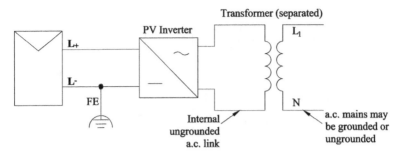

Figure 2.6 Functionally grounded PV system.

grounded system (Figure 2.5) has one conductor of the PV array intentionally connected to ground for purposes other than safety, and by means not necessarily sized in compliance with safety requirements [15].

The functional grounding may also be needed to improve the PV system performance (e.g., to prevent *Potential Induced Degradation* or "PID"). The functional grounding connection, typically located in the inverter, may be required by manufacturers as a condition for the proper operation of a specific PV module technology. The functional grounding may also be needed to improve the PV system performance (e.g., to prevent *Potential Induced Degradation* or "PID" [16]).

The functional connection to ground may be carried out through overcurrent devices, relays, or switching means that can allow the opening of the ground connection or its automatic disconnection in the event of ground faults.

Examples of non-functional earthing may include the use of *surge protection devices* or measuring devices of the array insulation resistance (impedance monitoring), which offer a very high impedance connection to ground.

According to IEC 60364, the functional grounding of a d.c. live conductor of the PV array is allowed only if the galvanic separation between d.c. and a.c. sides of the PV system is present (i.e., separation transformer) [17, 18].

In this type of system, a ground fault involving the positive conductor L_+ will impose a fault current that flows through the loop shown in Figure 2.7.

The fault current I_{PV} returns to the negative pole of the PV array by circulating through the functional grounding conductor (FE). I_{PV} is detected by either an overcurrent protective device in the functional ground conductor, which directly interrupts the fault current, or a current sensor K interfaced with the relay K_{PV}, which interrupts the fault current.

A ground fault may also involve the functionally grounded conductor L. (Figure 2.8).

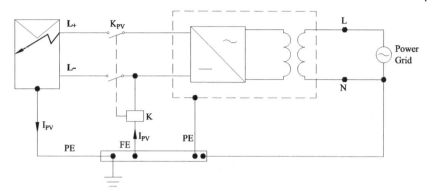

Figure 2.7 Functionally grounded system with faulty positive conductor.

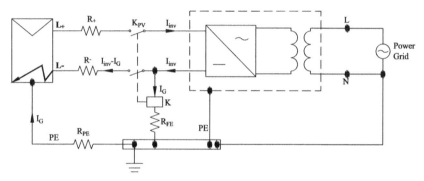

Figure 2.8 Functionally grounded system with faulty functionally grounded conductor.

The fault creates a fault current whose magnitude depends on the current divider formed between the resistance R^- of the functionally grounded conductor and the resistance of the path composed of the functional grounding conductor R_{FE}, and the protective conductor R_{PE} to the point of fault of the array.

R_{PE} and R_{FE} are typically very small if compared to R^-, therefore, a significant part of the inverter current I_{inv} flows into the fault as I_G. I_G can be detected by either an overcurrent protective device in the functional ground conductor, or a current sensor K interfaced with the relay K_{PV}.

If I_G is not large enough, or the inverter is not on (i.e., $I_{inv} = 0$), the fault current detection is not possible, and the fault may persist in the system.

To eliminate this problem, an Insulation Monitoring Device (IMD) to measure the PV array insulation resistance R_{iso} should be regularly used [19–21] The IMD may be permanently installed across the functionally grounded conductor and the ground busbar but made active for the measurement only after disconnecting the functional ground, and with K_{PV} closed (e.g., via interlocks).

The IMD would detect the insulation failure, regardless of the magnitude of the potential fault current.

2.6 Non-Ground-Referenced PV Systems

The non-ground-referenced PV system (or *ungrounded*) (Figure 2.9) is a PV array that has no conductors intentionally grounded, either directly or through the inverter.

All live parts are insulated from ground, whereas the exposed-conductive-parts are grounded. Frames and supporting structures of the PV generator must also be grounded.

In this type of system, the first ground fault on the PV array will produce a low magnitude fault current, as the return path is through the PV module insulation resistance. Thus, even though the detection of overcurrents in not essential for safety, the continuous insulation monitoring of the array is instead required.

Non-ground-referenced systems are defined as IT systems in IEC 60364-4-41.[3] The live parts of the PV systems are insulated from ground, whereas the exposed-conductive-parts (e.g., modules conductive frames) are grounded collectively, in groups or individually (Figure 2.10).

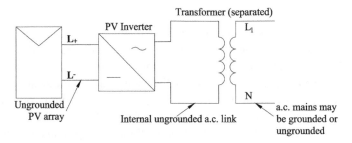

Figure 2.9 Non-ground-referenced PV system.

---

3 IEC 60364-4-41: "Low-voltage electrical installations – Part 4-41: Protection for safety – Protection against electric shock".

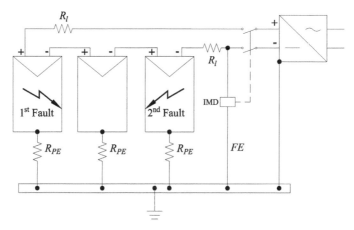

Figure 2.10   Second Fault in non-ground-referenced PV systems.

In the event of a single ground fault, the d.c. fault-current, in steady-state conditions, is of the same order of magnitude as the operation current, and the disconnection of the supply is not necessary for safety; thus, no limitation to the maximum value of the resistance of the ground-electrode for the PV system is required.

The first fault, due to the failure of the basic insulation of the PV generator, must be signaled to operators by means of a visual warning notified by the *Insulation Monitoring Device* (IMD). The IMD, installed across the negative and the ground terminals, permanently monitors the insulation R_{iso} between live parts and the ground, and when R_{iso} falls below a predetermined threshold, an alarm is output. The conductor to the IMD is a measuring connection and may be treated as functional ground connection (FE).

After the occurrence of a single ground fault, in the event of a second fault occurring on a different live conductor, a short-circuit current flows, driven by the pole-to-pole voltage (Figure 2.10). Per IEC standards, in d.c. ungrounded systems, the automatic disconnection of the supply on a second fault is not required if the condition given by Eq. 2.1 for collectively grounded exposed-conductive-parts is satisfied.

$$I_F = \frac{V}{2R_S} \geq I_a \tag{2.1}$$

V is the nominal d.c. voltage (in V) between line conductors; I_F is the fault-current circulating upon the second fault (in A); $R_S = R_l + R_{PE}$ is the resistance (in Ω) of the fault-loop comprising the line conductor and the protective conductor of the circuit; I_a is the current (in A) causing the operation of the

Table 2.1 Maximum disconnection times.

$120\,V_{dc} < V_o \leqslant 230\,V_{dc}$	$230\,V_{dc} < V_o \leqslant 400\,V_{dc}$	$V_o > 400\,V_{dc}$
5 s	0.4 s	0.1 s

protective device within the maximum disconnection times (in seconds) listed in Table 2.1, based on the d.c. nominal line-to-ground voltage V_o.

In d.c. systems with nominal voltage-to-ground less than 120 V (i.e., value conventionally deemed not hazardous), the disconnection of the supply upon the second fault is not required for protection against electric shock.

For greater voltages, the fulfillment of the condition expressed in Eq. 2.1, which is meant to facilitate the prompt operation of protective devices, is, however, based on voltage sources, which normally produce high fault-currents. PV arrays are instead current generators, and Eq. 2.1 may not always be satisfied; fault-currents, in fact, depend on the points of fault within the PV array, are of the same magnitude as the operating currents.

The magnitude of the second-fault current is low, therefore, the potential difference across two faulty PV panels, which is the voltage drop across the protective conductors interconnecting them, is generally below the threshold of danger of 120 V d.c., and the disconnection of the supply is consequently not required for safety.

However, at the second fault, inverter's product standards may require the operation of the main array d.c. switch, which disconnects the d.c. circuit to the inverter, preventing possible damages to its electronics. The response value R_{an} of the IMD must be consistent with the minimum insulation resistance R_{ISOm} of the PV generator occurring in normal operating conditions. To prevent nuisance disconnections, R_{an} should be chosen well below R_{ISOm} (e.g., $R_{an} \leq 0.6 R_{ISOm}$).

The loss, or lack, of the protective conductors between the faulty PV panels, is particularly dangerous, because a person may be subjected to the pole-to-pole voltage, which may be of a dangerous magnitude and might not be sensed by any protective device. Thus, the prompt repair of the first ground-fault plays an important role in the safety of ungrounded PV systems, because de-escalate the probability of the occurrence of the second fault.

2.7 Ground-Referenced PV Systems

In grounded PV systems, one pole of the PV generator is grounded, and all exposed-conductive-parts of the installation are connected by a protective conductor to the main ground busbar, which is locally earthed. For safety reasons,

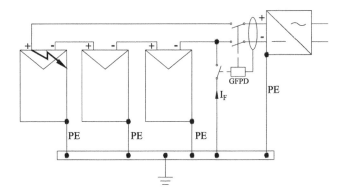

Figure 2.11 Ground Fault in PV systems.

it is recommended that the negative d.c. conductor be connected to the ground electrode at the d.c. side of inverter; this ground connection also reduces the electrochemical degradation of the electrode and other metallic parts.

The grounded system is defined as a TN system in IEC 60364-4-4 (Figure 2.11).

The characteristics of the protective devices and the circuit resistance in a TN d.c. system must satisfy the following requirement:

$$I_F = \frac{V_0}{R_S} \geq I_a \tag{2.2}$$

V_0 is the nominal d.c. line to ground voltage in volts (V); R_S is the resistance in ohms (Ω) of the fault loop comprising the source, the line conductor up to the point of the fault, and the protective conductor between the point of the fault and the source; I_a is the current (in A) causing the operation of the protective device within the maximum disconnection times listed in Table 2.1.

The magnitude of the fault current might not be high enough to satisfy Eq. 2.2 and guarantee the prompt disconnection of the supply.

A shock hazard current monitoring device is required to identify sudden changes in the residual or in the ground-fault current. Any current above the baseline steady-state current leaking in normal operating conditions will cause a *ground fault protection device* (GFPD) to operate.

The GFPD must feature low current thresholds and fast response, as listed in Table 2.2, according to IEC 63112.[4]

---

4 IEC 63112: "Safety, functionality and classification of Photovoltaic Earth Fault Protection (PV EFP) equipment." 2019.

Table 2.2 Residual or ground current thresholds and response time.

Residual or ground current threshold (mA)	Max response time (s)
30	0.3 s
60	0.15 s
90	0.04 s

In the event of a ground fault, the GFPD isolates the grounded pole of the PV array from earth, interrupts the fault current, and the PV generator goes into *floating* mode. After the resolution of the ground fault, the GFPD must be manually rearmed, and the grounding of the PV system is once again reestablished. In addition, product standards may also require the immediate disconnection of the inverter from the PV array to prevent damages; therefore, in the event of ground faults, the inverter shutdown may also occur, similar to the case of low values of the insulation to ground of the PV array.

2.8 Fire Hazard in Ground-Referenced PV Systems

Ground-fault currents in ground-referenced PV systems may cause major fire accidents if excessive ground currents are allowed to flow on conductive parts that are not intended to carry any current. The fire hazard current monitoring may be implemented by an overcurrent protective device installed in the grounding conductor, which detects abnormal ground fault currents circulating through it (Figure 2.12). Ground-fault currents in ground-referenced PV systems may cause major fire accidents [22], if excessive ground currents are allowed to flow on conductive parts that are not intended to carry any current [23, 24]. The fire hazard current monitoring may be implemented by an overcurrent protective device installed in the grounding conductor, which detects abnormal ground fault currents circulating through it (Figure 2.11).

The overcurrent device, which may be referred to as *Ground Fault Detection Interrupter* (GFDI) [25], interrupts the excessive current I_F in the grounding conductor, therefore limiting the ground fault energy and reducing the fire risk. The PV generator becomes ungrounded and is not functioning in the proper design operating conditions. For this reason, the GFDI is equipped with auxiliary contacts to a relay that shuts down the inverter, preventing abnormal operations.

The rated current of the overcurrent protection in the grounding conductor, for various PV array power ratings, are provided in IEC 63112, and in UL 1741[5] (Table 2.3).

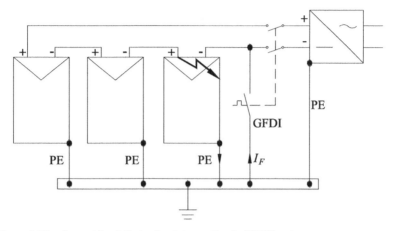

Figure 2.12 Ground Fault Detection Interruption in TN PV systems.

Table 2.3 Rated current of overcurrent protection in the functional grounding conductor.

Max PV array DC rating at STC[6] (kW)	Rated current (A)
≤ 25	1
> 25 and ≤ 50	2
> 50 and ≤ 100	3
> 100 and ≤ 250	4
> 250	5

5 UL 1741: "Standard for Inverters, Converters, Controllers and Interconnection System Equipment for Use With Distributed Energy Resources." 2010.
6 STC (Standard Test Conditions) refers to the fixed set of laboratory conditions under which every solar module is tested: PV module temperature 25°C, irradiance 1 kW/m², AM 1.5.

A defect in the insulation of the grounded conductor may cause part of the operating current to flow back to the PV generator via the GFDI (which is normally closed) (Figure 2.13).

This current I', proportional to the irradiance over the PV array, will flow through an unintended path and may also cause the tripping of the GFDI. The GFDI does not ensure personnel protection, given the high values of rating currents, but only system protection.

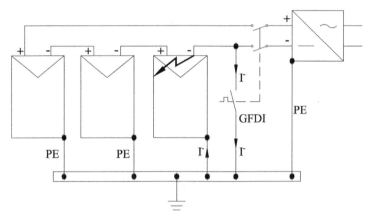

Figure 2.13   Defect in the insulation of the grounded conductor.

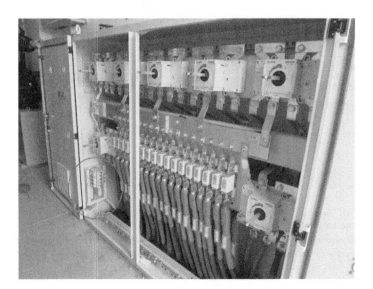

Figure 2.14   GFDI installed in combiner box.

The GFDI may be part of the inverter or be installed, as a separate unit, in the combiner box, which collects all d.c. inputs from the strings of the PV array (Figure 2.14).

The GFDI is always interlocked with the inverter, which shuts down upon the GFDI operation.

2.9 Faults at Loads Downstream the PV Inverter in Ground-referenced PV Systems

The point of common coupling L-N between the PV systems and the power grid must be upstream of the load protective devices (Figure 2.15).

In the case of TN systems, ground-faults downstream the point of common coupling will be sensed by protective devices, which will detect both contributions to the fault: I_{PV} impressed by the PV generator via the inverter, and I_G, impressed by the power grid. The protective device, which may consist of either a *Residual Current Device* (RCD) [26] or an overcurrent device, or both, can disconnect the supply and protect persons against indirect contact. The RCD is a device designed to make, carry and break currents under normal service conditions and to cause the opening of the contacts when the *residual current* attains a specific threshold. The *residual current* is the RMS value of the vector sum of the instantaneous values of the currents flowing through the main circuit of the residual current device.

Should a ground fault occur within the inverter, the fault could only be cleared by a dedicated protective device (e.g., a B type RCD) installed immediately downstream of it, at the a.c. terminals (Figure 2.16).

The RCD disconnects the line, and the inverter, sensing the lack of voltage at its terminals, goes into stand-by mode. Type B RCDs can detect a.c., pulsating

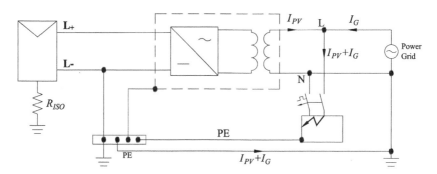

Figure 2.15 Faults downstream the PV inverter in ground-referenced PV systems.

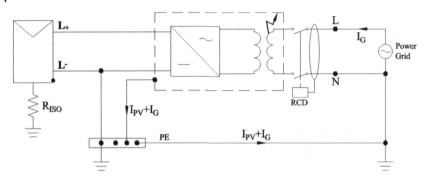

Figure 2.16 Faults within the inverter.

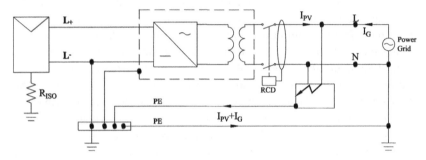

Figure 2.17 Fault occurring at load between inverter and point of common coupling.

d.c., composite of multi-frequency and d.c. residual currents, such as those possibly output by high-frequency transformers.

The RCD in question cannot sense a ground-fault occurring within any exposed-conductive-part connected between the RCD and the point of common coupling L-N (Figure 2.17).

The RCD, in fact, cannot sense any unbalance in the current passing through the toroid; thus, such exposed-conductive-parts must be protected by a dedicated protective device, or must not be connected between the inverter and the point of common coupling L-N.

2.10 Non-Electrically Separated PV System

The non-electrically-separated PV system (Figure 2.18) is a PV array whose live conductors are not grounded but is connected to an intentionally ground-referenced a.c. system through the non-separated inverter (i.e., *transformerless inverters* [27]).

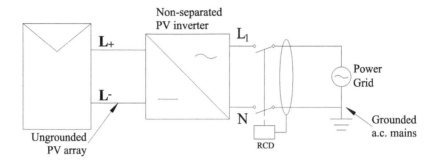

Figure 2.18 Non- electrically-separated PV system.

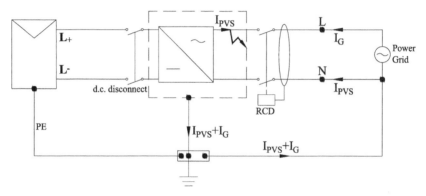

Figure 2.19 Transformerless PV inverter and ground fault on the a.c. side.

When the inverter is connected (e.g., all system disconnects are closed), the entire system, including the PV array, becomes ground-referenced. The non-separated inverter typically includes switching devices that disconnect it from the PV array and the a.c. mains in the events of faults.

In a non-separated system, either the positive or the negative conductors of the PV array should not be functionally grounded to prevent normal operating currents from flowing through the actual ground.

As already discussed, PV inverters have been traditionally equipped with transformers that provide galvanic isolation from the power grid and isolation between the PV system and the grid grounding system. However, the transformer reduces the PV inverter efficiency and increases its size and costs. For this reason, inverters can be *transformerless* (Figure 2.19).

If there is a ground fault on the a.c. side (Figure 2.17), the fault current, composed of the power grid current I_G and the a.c. PV system current I_{PVS}, causes the operation of the RCD installed on the a.c. side of the inverter, which senses

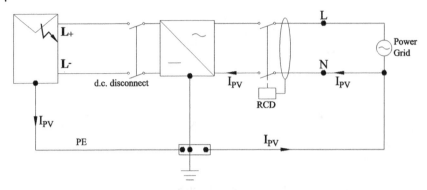

Figure 2.20 Transformerless PV inverter and ground fault on the d.c. side.

a current unbalance. Once the RCD disconnects the power grid from the PV system, the inverter, thanks to the *anti-islanding* feature [28–31] based on the lack of grid voltage at its output, shifts to standby, thereby completely clearing the ground fault. However, even in this situation, the solar radiation keeps energized the array conductors.

If there is a fault on the positive conductor of the array (Figure 2.20), the fault current I_{PV} will return to the negative pole through the a.c. grounded neutral, passing through the RCD, causing a current unbalance.

If the RCD is sensitive to d.c. currents (i.e., type B), the current unbalance is detected, and the a.c. supply is disconnected. The ground fault is then completely cleared when the inverter, sensing the lack of power grid voltage at its a.c. terminals, will shut down.

2.11   PV Systems Wiring Methods and Safety

Wiring methods used in PV systems include special cables, connectors, and other methods specifically identified for use in PV systems; standard types of conductors, raceways, and fittings are also normally employed.

d.c. cables of photovoltaic installations must satisfy more stringent safety requirements due to the expected harsh conditions of use, such as exposure to UV, rain, and high temperatures. PV modules may operate at temperatures of the order of 40°C above ambient temperature, thus the insulation of cables installed in contact or near PV modules, such as *string* cables, must be accordingly rated.

PV modules submerged in the body of water have been considered to reduce operating temperatures and increase efficiency, as well as reduce environmental impact and cleaning issues. A shallow immersion of the module (1 or 2 cm) in water at a temperature of around 15°C allows an increase in efficiency, which greatly compensates the loss due to the sun radiation absorption of the water.[7]

PV string cables connect the modules in a PV string, and/or the string to a junction box or to the d.c. terminals of the inverter. PV string cables, PV array cables and PV d.c. main cables must be selected and installed to minimize the risk of ground faults and short-circuits. This may be achieved by employing *photovoltaic (PV) wires*, that is, double insulated single-core cables featuring a layer of insulation and a protective non-metallic sheath: these two layers fulfill the requirements for both basic and fault protection. The sheath does not provide for an electrical insulation, even if made of an insulator, but only for mechanical protection during the installation.

Cables can be considered Class II if they have a non-metallic sheath and an insulation of a higher rating than the operating voltage of the electrical system. As an example, for a system with an operating voltage of 230/400 V, cables with sheath with ratings 300/500 V, 450/750 V or 0.6/1 kV are considered of Class II.

The basic insulation of PV cables must be rated in accordance with the maximum open-circuit voltage of the array $V_{oc\ max}$, which must be calculated at the record-low ambient temperature at the site, per IEC 61829.[8]

The determination of $V_{oc\ max}$ is, however, possible only if the open-circuit voltage temperature coefficients are included in the PV module manufacturer's datasheet; if this is not the case, a simplified approach is presented in the standard IEC TS 62257-7-1.[9] The standard indicates that the insulation of PV cables should have a voltage rating of at least 1.2 times nV_{oc}, where n is the number of the series-connected PV modules in any PV string, and V_{oc} is the open-circuit voltage of the single module in standard test conditions (STC), which is an information typically provided in the datasheet. The standard test conditions used for the testing and rating of PV modules consist of cell

7 R. Cazzaniga *et al.*: "*Floating photovoltaic plants: Performance analysis and design solutions*," Renewable and Sustainable Energy Reviews, Volume 81, Part 2, January 2018, Pages 1730-1741.

8 IEC 61829: "*Photovoltaic (PV) Array - On-Site Measurement of Current-Voltage Characteristics*".

9 IEC TS 62257-7-1: "*Recommendations for small renewable energy and hybrid systems for rural electrification - Part 7-1: Generators - Photovoltaic generators*".

temperature of 25ºC, irradiance in the plane of the PV cell or module of 1000 W/m^2 and light spectrum corresponding to an atmospheric air mass of 1.5.

Cables and PV equipment may have a nominal d.c. voltage rating of 1.5 kV, both between conductors and between conductors and ground, and per EN 50618[10] the maximum permitted operating d.c. voltage of the PV system must not exceed 1.8 kV.

In the U.S., the National Electrical Code (NEC) also permits, as an alternative to the PV wire, the USE-2 wire (i.e., *Underground Service Entrance*), which is also tested for exposure to sunlight. The NEC also prescribes that PV circuits operating at voltages greater than 30 V installed in readily accessible locations, for physical protection and to reduce electrical hazards, must be Type MC (Metal Clad) Cable, featuring a galvanized interlocking steel armor, or installed in a raceway.

2.12   d.c. Currents and Safety

PV installations further challenge our electrical safety: a.c. voltages and d.c. voltages (possibly in excess of 600 V) coexist in the same system; PV generators cannot be turned off during maintenance.

As previously discussed, for shock durations longer than the cardiac cycle, the threshold of fibrillation for the d.c. current is higher than that for the a.c. current, therefore, the former is generally less dangerous. Experiments on animals, as well as data derived from electrical accidents, have also demonstrated that the threshold of fibrillation for a d.c. downward current, is about twice as high as for an upward current. Thus, grounded PV systems should have the negative conductor earthed, so that a contact with a live part, at positive potential, would expose persons to a downward flow of current, with less risk of ventricular fibrillation.

Conventional time/current curves describing the effects of d.c. currents on persons for shock durations of 10 ms and longer, and for a longitudinal (i.e. hand to feet) upward current path, are provided in IEC 60479-1[11] (Figure 2.21). The description of the time/current zones are shown in Table 2.4.

---

10 EN 50618: *"Electric Cables for Photovoltaic Systems"*.

11 IEC/TS 60479-1: *"Effects of Current on Human Beings and Livestock - Part 1: General Aspects"*.

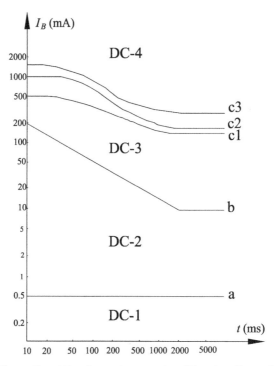

Figure 2.21 Conventional time/current curves describing the effects of d.c. currents.

Table 2.4 Description of the time/current zones for d.c. currents.

Zones	Boundaries	Physiological effects
DC-1	2 mA to curve a	Slight pricking sensation possible when making, breaking, or rapidly altering current flow.
DC-2	2 mA up to curve b	Likely involuntary muscular contractions, but usually no harmful electrical physiological effects.
DC-3	Curve b and above	Strong involuntary muscular reactions and reversible disturbances in the heart, increasing with current magnitude and time. Usually no organic damage.
DC-4	Above curve c_1	Cardiac arrest, breathing arrest, burns, or other cellular damage.
	c_1–c_2	Probability of ventricular fibrillation increasing up to about 5 %.
	c_1–c_3	Probability of ventricular fibrillation up to about 50 %
	Beyond curve c_3	Probability of ventricular fibrillation above 50 %

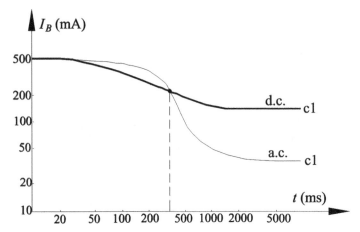

Figure 2.22 Comparison between a.c. and d.c. fibrillation curves.

It is important to note that according to the conventional time/current curves of Figure 2.21, curve c_1 is the threshold of inception of ventricular fibrillation; above curve c_1 the probability of ventricular fibrillation increases with current magnitude and time.

A comparison with the time/current curve c_1 for a.c. currents provided in the IEC 60479-1 standard shows that for contact times below approximately 330 ms, the body current that may trigger the ventricular fibrillation is lower for the direct current (Figure 2.22); thus, in that time-domain, the d.c. current is more dangerous than the a.c current.

A relevant technical standard that can be used to define maximum permissible touch voltages U_{tM} in PV systems as a function of the contact duration t is the EN 50122-1.[12] This standard sets permissible voltages for short-term contact durations in d.c. traction systems to lower values than those for a.c. traction systems (Figure 2.23).

For contact durations less than approximately 330 ms, the risk of inception of heart fibrillation is higher in d.c. systems, hence lower values of U_{tM} are accordingly established.

For example, in correspondence with $t = 0.2$ s, U_{tM} for a.c. voltages is 645 V, whereas U_{tM} in d.c. voltages is 520 V.

The above touch voltage limits have been established by the standard considering in series to the human body two additional resistances R_{a1} and R_{a2}: R_{a1}

---

12 EN 50122-1: "Railway applications - Fixed installations - Electrical safety, earthing and the return circuit - Part 1: Protective provisions against electric shock".

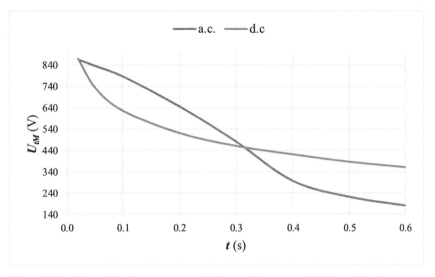

Figure 2.23 Comparison between a.c. and d.c. maximum permissible touch voltages.

is the resistance of the person to ground in the conservative case of the absence of floors, which is equal to $1.5\rho_s$, where ρ_s is the standing surface resistivity in Ωm (150 Ω in the diagram of Figure 2.13); R_{a2} is the resistance provided by old and wet shoes (e.g., workers at site), assumed to be 1 kΩ.

The risk of ventricular fibrillation for the pathway hand-to-hand is less dangerous than the path hand to feet, even when R_{a2} is included, and this is regardless of the direction of the d.c. current.

2.13 Electrical Safety of PV Systems

When the sun is shining, PV generators are always *live*, even if the array d.c. disconnect is open, and equipment have energized parts that may be touched during installation or maintenance. After sunset, experiments revealed that values of open-circuit voltages and short-circuit currents of residential PV modules [32] illuminated by the high-intensity flood lamps used by fire departments during nighttime firefighting operations, were hazardous.[13]

---

13 R. Backstrom, D.A. Dini: "UL Firefighter Safety and Photovoltaic Installations Research Project". (2011).

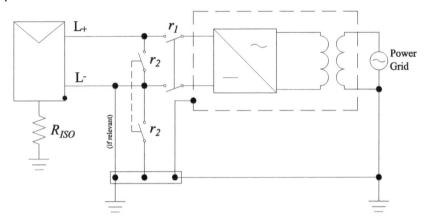

Figure 2.24 Putting in safety PV generators.

Electrical work on PV systems should only be conducted by qualified workers with sufficient technical knowledge and experience in dealing with electricity [33].

Electrical work is defined as *work on or near an electrical installation where an electrical hazard is present*. Electrical work on PV systems may include testing and measuring, repairing, replacing, modifying, extending, erection and inspection.

During the erection of PV arrays, the terminals of each PV module, or of a string, may be short-circuited to cancel potential differences. The short-circuit does not damage the PV generators, as the resulting current is of the same magnitude as the operating current.

To allow workers to maintain PV generators by operating in de-energized conditions (i.e., *dead* working), a shutdown procedure consisting of short-circuiting and grounding the live conductors at the inverter input terminals may be implemented (Figure 2.24).

To prevent damages to the inverter, an interlock must prevent the contacts r_2 from closing if the contacts r_1 of the main array switch are closed. As per the EN 50110-1,[14] this shut-down procedure allows the *dead* working, as opposed to the *live* working or *work in the vicinity of live parts* methods; however, the *dead* working is predicated on the worker verifying that the installation is, in fact, de-energized and secured against the re-energization by means of lock-out

---

14 EN 50110-1: "*Operation of electrical installations - Part 1: General requirements*".

and tag-out of the d.c. and a.c. disconnects, shut down the inverter and opening protective devices in the combiner box.

If the shutdown procedure cannot be implemented, the *live* working method must be followed. As per EN 50110-1, the *live* working zone is a space around energized parts of the PV installation in which the reduction of electrical risk is achieved by always employing two safety barriers: should one barrier fail there will still be one barrier providing the worker with complete safety. The safety barriers may consist of personal protective equipment, such as electrical insulating gloves, and insulated tools for *live* working.

For system voltages less than 1 kV, the distance from the energized parts to the outer boundary of a *live* working zone is considered as zero; thus, *live* work is defined as an activity in which the worker may contact the energized part with parts of their body or with tools, devices or equipment.

Assuming all safety measures are effective, the *live* work method provides equivalent levels of safety as the *dead* work method.

2.14 Rapid-Shutdown of PV Arrays on Buildings

During emergencies, such as building fires, to guarantee the safety of first responders, all sources of power must be shut off [34]. Before spraying water, or climbing on the roof, firefighters should also be able to deenergize the PV arrays. If the sun shines, or at night when apparatus-mounted scene lighting is employed, PV strings may generate an electrical hazard; the d.c. wires between the roof and the inverter may still be at a dangerous voltage, even if the main a.c. breaker at the service panel is shut off.

To protect personnel during operation on rooftops, section 690.12 of the U.S. National Electrical Code (NEC) has introduced the safety feature of the *rapid shutdown* of PV systems. The primary objective of the rapid shutdown functionality is to quickly reduce the voltage of the d.c. circuits to safe levels, by means of a quick and easy-to-activate procedure.

The NEC introduces the term *array boundary* defined as the region extending 30 cm from the PV array in all directions (Figure 2.25).

d.c. conductors located outside the array boundary, or more than 1 m from the point of entry inside a building, must attain a voltage not exceeding 30 V within 30 s of the rapid shutdown initiation.

d.c. conductors located inside the boundary, or not more than 1 m from the point of penetration of the surface of the building, must attain a voltage not exceeding 80 V within 30 s of the rapid shutdown initiation.

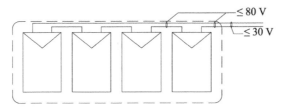

Figure 2.25 Array boundary and d.c. voltage requirements.

As earlier mentioned, 80 Vdc is below the conventional threshold of danger of 120 Vdc identified as such by international standards. However, OSHA does not distinguish between a.c and d.c voltages and considers all voltages of 50 V or above to be hazardous.

To comply with the 80 V inside-the-boundary requirement, it is not enough to disconnect the strings connected in parallel with each other. Every PV module of the array will need to be isolated from each other so that they cannot add up their output voltages within the string.

The rapid-shutdown feature may be accomplished with module-level electronics, such as optimizers and micro-inverters, which, upon the shut off of the main a.c. breaker at the service panel, within a few cycles reduce the PV module voltage to zero or to safe values.

String inverter configurations may be retrofitted with switching equipment and communication units that can disconnect from each other the series connected modules.

2.15 Hazard and Risk

The safety of personnel must be based on the identification of the hazards, the assessment of the related risks, and the implementation of safety measures to lower the risk below an acceptable threshold.

The risk exposure R is defined as the probability of an unfortunate event occurring, multiplied by the potential impact or damage incurred by the event, which can be depicted by iso-risk curves (i.e., curves at constant risk) (Figure 2.26).

The iso-risk curves identify the same risk given by different combinations of probability and damage. The risk can be optimally reduced, thus going from curve R_2 to curve R_1 ($R_2 < R_1$), by increasing prevention (i.e., reducing the probability of unfortunate events) or by increasing protection, (i.e., reducing the magnitude of damage).

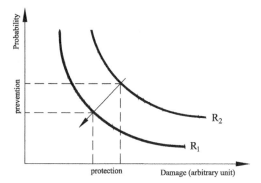

Figure 2.26 Iso-risk curves.

Controlling the hazard at its source is the best way to protect personnel. According to the U.S. OSHA,[15] the two key factors that would determine whether a control to reduce the risk exposure is practicable are its technological and economic feasibilities. The cost of implementing engineering and administrative controls, if available, must not threaten the employer's ability to remain in business. If these two factors cannot be satisfied, alternative approach must be implemented by the employer (e.g., providing personal protective equipment).

References

1 Cabrera-Tobar, A., Bullich-Massagué, E., Aragüés-Peñalba, M., and Gomis-Bellmunt, O. (2016). Topologies for large scale photovoltaic power plants. *Renewable and Sustainable Energy Reviews* 59: 309–319.

2 Joshi, A.S., Dincer, I., and Reddy, B.V. (October 2009). Performance analysis of photovoltaic systems: A review. *Renewable and Sustainable Energy Reviews* 13 (8): 1884–1897.

3 Teodorescu, R., Liserre, M., and Rodríguez, P. (2010). Photovoltaic inverter structures. In: *Grid Converters for Photovoltaic and Wind Power Systems*, 5–29. John Wiley & Sons, Ltd.

15 The Occupational Safety and Health Administration is an agency of the United States Department of Labor, with the mission to "assure safe and healthy working conditions for working men and women by setting and enforcing standards and by providing training, outreach, education and assistance".

4 Kouro, S., Leon, J.I., Vinnikov, D., and Franquelo, L.G. (2015). Grid-connected photovoltaic systems: An overview of recent research and emerging PV converter technology. *IEEE Industrial Electronics Magazine* 9 (1): 47–61.

5 Reshma Gopi, R. and Sreejith, S. (2018). Converter topologies in photovoltaic applications – A review. *Renewable and Sustainable Energy Reviews* 94: 1–14.

6 Araneo, R., Lammens, S., Grossi, M., and Bertone, S. (2009). EMC issues in high-power grid-connected photovoltaic plants. *IEEE Transactions on Electromagnetic Compatibility* 51 (3): PART 2.

7 Fekete, K., Klaic, Z., and Majdandzic, L. (2012). Expansion of the residential photovoltaic systems and its harmonic impact on the distribution grid. *Renewable Energy* 43: 140–148.

8 Vinayagam, A., Aziz, A., Pm, B., Chandran, J., Veerasamy, V., and Gargoom, A. (2019). Harmonics assessment and mitigation in a photovoltaic integrated network. *Sustainable Energy, Grids and Networks* 20: 100264.

9 Anand, S., Gundlapalli, S.K., and Fernandes, B.G. (2014). Transformer-less grid feeding current source inverter for solar photovoltaic system. *IEEE Transactions on Industrial Electronics* 61 (10): 5334–5344.

10 Carrasco, J.M., et al. (2006). Power-electronic systems for the grid integration of renewable energy sources: A survey. *IEEE Transactions on Industrial Electronics* 53 (4): 1002–1016.

11 Araneo, R. and Mitolo, M. (2019). On the insulation resistance in high-power free-field grid-connected photovoltaic plants. *Proceedings - 2019 IEEE International Conference on Environment and Electrical Engineering and 2019 IEEE Industrial and Commercial Power Systems Europe, EEEIC/I and CPS Europe 2019*.

12 Lorenzo, G.D., Araneo, R., Mitolo, M., Niccolai, A., and Grimaccia, F. (2020). Review of O M practices in PV plants: Failures, solutions, remote control, and monitoring tools. *IEEE Journal of Photovoltaics* 10 (4): 914–926.

13 Flicker, J. and Johnson, J. (2016). Photovoltaic ground fault detection recommendations for array safety and operation. *Solar Energy* 140: 34–50.

14 Albers, M.J. and Ball, G. (2015). Comparative evaluation of DC fault-mitigation techniques in large PV systems. *IEEE Journal of Photovoltaics* 5 (4): 1169–1174.

15 Falvo, M.C. and Capparella, S. (2015). Safety issues in PV systems: Design choices for a secure fault detection and for preventing fire risk. *Case Studies in Fire Safety* 3: 1–16.

16 Hacke, P. et al. (2011). System voltage potential-induced degradation mechanisms in PV modules and methods for test. *2011 37th IEEE Photovoltaic Specialists Conference*, 814–820.

17 Pillai, D.S. and Natarajan, R. (2019). A compatibility analysis on NEC, IEC, and UL standards for protection against line–line and line–ground faults in PV arrays. *IEEE Journal of Photovoltaics* 9 (3): 864–871.

18 Wu, Y., Lin, J., and Lin, H. (2017). Standards and guidelines for grid-connected photovoltaic generation systems: A review and comparison. *IEEE Transactions on Industry Applications* 53 (4): 3205–3216.

19 Hernández, J.C., Vidal, P.G., and Medina, A. (2010). Characterization of the insulation and leakage currents of PV generators: Relevance for human safety. *Renewable Energy* 35 (3): 593–601.

20 Hernández, J.C. and Vidal, P.G. (2009). Guidelines for protection against electric shock in PV generators. *IEEE Transactions on Energy Conversion* 24 (1): 274–282.

21 Pillai, D.S. and Rajasekar, N. (2018). A comprehensive review on protection challenges and fault diagnosis in PV systems. *Renewable and Sustainable Energy Reviews* 91: 18–40.

22 Manzini, G., Gramazio, P., Guastella, S., Liciotti, C., and Baffoni, G.L. (2015). The fire risk in photovoltaic installations – Checking the PV modules safety in case of fire. *Energy Procedia* 81: 665–672.

23 Mellit, A., Tina, G.M., and Kalogirou, S.A. (2018). Fault detection and diagnosis methods for photovoltaic systems: A review. *Renewable and Sustainable Energy Reviews* 91: 1–17.

24 Pillai, D.S., Blaabjerg, F., and Rajasekar, N. (2019). A comparative evaluation of advanced fault detection approaches for PV systems. *IEEE Journal of Photovoltaics* 9 (2): 513–527.

25 Dini, D.A., Brazis, P.W., and Yen, K. (2011). Development of arc-fault circuit-interrupter requirements for photovoltaic systems. *2011 37th IEEE Photovoltaic Specialists Conference*, 1790–1794.

26 Pillai, D.S., Ram, J.P., Rajasekar, N., Mahmud, A., Yang, Y., and Blaabjerg, F. (2019). Extended analysis on line-line and line-ground faults in PV arrays and a compatibility study on latest NEC protection standards. *Energy Conversion and Management* 196: 988–1001.

27 Saridakis, S., Koutroulis, E., and Blaabjerg, F. (2013). Optimal design of modern transformerless PV inverter topologies. *IEEE Transactions on Energy Conversion* 28 (2): 394–404.

28 Velasco, D., Trujillo, C.L., Garcerá, G., and Figueres, E. (2010). Review of anti-islanding techniques in distributed generators. *Renewable and Sustainable Energy Reviews* 14 (6): 1608–1614.

29 De Mango, F., Liserre, M., Dell'Aquila, A., and Pigazo, A. (2006). Overview of anti-islanding algorithms for PV systems. Part I: Passive methods. *2006 12th International Power Electronics and Motion Control Conference*, 1878–1883.

30 De Mango, F., Liserre, M., and Dell'Aquila, A. (2006). Overview of anti-islanding algorithms for PV systems. Part II: Active methods. *2006 12th International Power Electronics and Motion Control Conference*, 1884–1889.

31 Yu, B., Matsui, M., and Yu, G. (2010). A review of current anti-islanding methods for photovoltaic power system. *Solar Energy* 84 (5): 745–754.

32 Laukamp, H. and Bopp, G. (July 1996). Residential PV systems—electrical safety issues and installation guidelines. *Progress in Photovoltaics: Research and Applications* 4 (4): 307–314.

33 Fthenakis, V.M. and Moskowitz, P.D. (Jan. 2000). Photovoltaics: Environmental, health and safety issues and perspectives. *Progress in Photovoltaics: Research and Applications* 8 (1): 27–38.

34 Spataru, S., Sera, D., Blaabjerg, F., Mathe, L., and Kerekes, T. (2013). Firefighter safety for PV systems: Overview of future requirements and protection systems. *2013 IEEE Energy Conversion Congress and Exposition*, 4468–4475.

3

Grounding and Bonding

CONTENTS

Mathematics is a game played according to certain simple rules with meaningless marks on paper.

David Hilbert

3.1 Introduction

The proper design of grounding systems is key to the effective and safe operation of renewable power plants. The first applications of grounding in electrical engineering can be dated back to the early 1800s (i.e., first applications of the telegraph); however, the novel grounding practice for renewable energies sources (RESs) needs further clarifications to prevent common misunderstanding and misconceptions. Particularly, the increase in complexity of RESs, which are often operated at various a.c. and d.c. voltage levels and frequencies, with the integration of a large number of electrical and electronic devices, calls

Electrical Safety Engineering of Renewable Energy Systems. First Edition. Rodolfo Araneo and Massimo Mitolo.
© 2022 by The Institute of Electrical and Electronics Engineers, Inc. Published 2022 by John Wiley & Sons, Inc.

for a better understanding of grounding based on the physical phenomena behind it.

Let us start from the definition of the terms *"grounding"* (or *"ground"*), which may be misused by practitioners. The term grounding (G) is adopted in North America, whereas in other English-speaking countries, the predominant term is *Earth* (E), translated into *Terre* in the standard of the International Electrotechnical Commission IEC 60364.

Generally speaking, *electrical grounding* refers to the process of joining parts of an electrical/electronic system to a reference conductive body, by means of a low-impedance connection. This statement is supported by the definition of ground found in IEEE Std. 1100 [1], which defines ground as a *conducting connection, whether intentional or accidental, by which an electric circuit or equipment is connected to the earth or to some conducting body of relatively large extent that serves in place of the earth. Note: Grounds are used for establishing and maintaining the potential of the earth (or of the conducting body), or approximately that potential, on conductors connected to it and for conducting ground currents to and from earth (or the conducting body).*

In power systems engineering, the term grounding indicates a connection to earth and may be interchangeably used with the word *earthing* [2, 3]. The earth is used as a reference and is considered to have zero voltage or potential.

Grounding may be implemented for two main purposes: (i) safety and (ii) functionality.

In *safety grounding*, the first and foremost goal is to provide measure(s) of safety to preserve life and property against hazards (e.g., electric shocks and fires) in the event of faults, transient overvoltages due to lightning, switching operations or other surge events, or accidental contact with higher-voltage systems.

The ground connection must redirect electric currents to ground without any operating electrical limits for the equipment being exceeded, or adversely affecting the continuity of the electrical service. Through proper design, it must always be assured that a person in the vicinity of the grounded equipment is not exposed to the danger of electric shock. Safety grounding design is based on low-frequency operating voltages and currents, such as 50 or 60 Hz, or even d.c.

Grounding is related both to *equipment* [4] as well as to *power electrical systems*.

The equipment grounding is the process of connecting through a conductor, referred to as the *Equipment-Grounding Conductor* (EGC) in the National Electrical Code (NEC) in the United States, or *Protective Earth* (PE) conductor [5] in IEC Standards, to an equipment frame, cabinet, or other enclosure, to the grounding system, with the purpose of preventing electric shock. The objective of the equipment grounding is to provide a low-impedance path to the fault current to facilitate the operation of overcurrent protection devices (OCPD), such as fuses, circuit breakers, relays, or residual current devices (i.e., RCDs, GFCIs), in the event of ground-faults.

System grounding is the process of intentionally making a neutral-to-ground connection at one point within the system by means of a conductor, referred to as the *grounding electrode conductor* in the NEC. Objectives of the system grounding are to stabilize the system voltage-to-earth during both normal operating conditions and fault conditions. System grounding can be of four different types: ungrounded systems, resistance grounding, reactance grounding, and solid grounding. In the United States, the NEC prescribes that a premises wiring system supplied by a grounded a.c. low-voltage service must have the grounding electrode conductor connected to the grounded service conductor (i.e., the neutral wire), to establish a neutral-to-ground connection at the service entrance panel.

Power systems designs according to IEC 60364 are based on five earthing types: TN-S, TN-C-S, TN-C, TT, and IT. Ungrounded (i.e., IT) RESs are common in Europe, and do not require the disconnection of the supply for safety reasons at the first phase-to-ground fault. However, many investigations have revealed that when ground faults occur in one phase, the unfaulted phases may experience steady-state or transient phase-to-ground overvoltages, which may result in insulation failures, and phase-to-phase faults.

Besides safety grounding, IEC 60050 introduces the *functional grounding*, which is the grounding for purposes other than electrical safety. This definition may include alternative terms that are often used by electrical engineers involved in SI/EMC [7], such as *signal grounding, EMI grounding, reference grounding*, etc. Functional grounding is primarily employed to allow electrical installations to operate in an electromagnetically stable fashion. It allows undisturbed operation of sensitive electronic devices, especially in an electromagnetic polluted environment and is essential to ensure an acceptable performance of electrical and electronic equipment. Functional grounding operates with frequencies ranging from d.c. to the highest frequency present in the signal spectrum.

In the case of high-frequency signals travelling in printed circuit boards, the ground may serve as a current return path for signals referenced to it. The earth can also act as the return path for normal operating currents at 50–60 Hz in the *Single Wire Earth Return* (SWER) transmission system adopted in unpopulated areas of some countries (e.g., India, New Zealand, Australia).

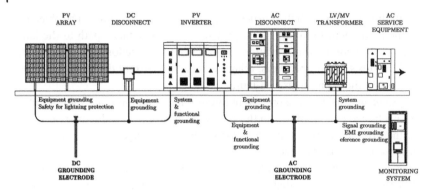

Figure 3.1 Schematics of the grounding system of a PV installation.

The novice practitioner may be surprised to realize that in renewable energy installations, which may be perceived as less complex than conventional power plants, all the above groundings may be encountered [8–11].

To make more challenging the understanding of grounding is that *bonding* can simultaneously play other crucial roles.

For instance, let us consider a photovoltaic installation, as shown in Figure 3.1.

- We have grounding connection on both the a.c. side (e.g., switchboard frames and cable trays) and the d.c. side (e.g., solar racks, support structures);
- We have a system grounding on the a.c. side, if the source is operated as solidly grounded.
- We may have a system and functional grounding on the d.c. side, which is typical of the United States, where either the positive or the negative d.c. wire is grounded; or the center tap of *bipolar* PV arrays[1] is grounded. The functional grounding is effective in reducing, for example, the *Potential-Induced Degradation* (PID) of certain types of thin-film PV panels.
- We may have functional grounding in inverters, Battery Energy Storage Systems (BESS), but also in communication systems used for monitoring, control, and security of alternative energy systems, such as *modbus* systems, whose cable shields must be properly grounded to reduce electrostatic and electromagnetic interferences. Enclosures and additional shielding of wireless and satellite internet equipment and closed-circuit television (CCTV) are also normally connected to the functional earth.

---

1 A bipolar PV array has two outputs, each having opposite polarity to a common reference point or *center tap*.

- We may have a functional grounding for double-insulated PV arrays (i.e., Class II) since their metal frames should be connected to the ground to allow the operation of insulation monitoring device (as per IEC Std. 61730-1).

The *temporary protective grounding* [12] provides a critical safety function when working on electrically isolated, or de-energized, a.c. and d.c. installations and must be installed between the ungrounded conductor(s) and the earth. To establish electrically safe work conditions, a *lockout & tagout* procedure must always be carried out on normally live equipment, which must then be properly tested to confirm the absence of voltages. The temporary protective grounding must not be confused with the *static grounding*, which is only used to prevent isolated-from-ground objects from accruing a static charge.

Temporary protective grounds prevent the effects on workers of the accidental reenergization of circuits or equipment by maintaining the voltage at the work location within a safe level until the circuit is disconnected. The temporary protective grounding protects against any accidental reenergization, for instance, caused by induced voltages or fault-current feed-over from adjacent energized lines, lightning strikes, switching equipment malfunction, or human error.

3.2 Basic Concepts of Grounding Systems: The Ground Rod

The resistance to earth of ground-electrodes plays a key role in the successful operation of any electrical system, especially in the renewable energy sector, where high technology equipment have stringent grounding requirements. The importance of the understanding of the behavior of the grounding system is summarized by the statement: "*To protect what is above the ground you need to know what is in the ground.*"

A grounding system is generally made of both horizontal and vertical electrodes buried in the soil. The head of ground-rods may be located at any depth below ground; however, for horizontal electrodes, a minimum burial depth of 0.5 m to 1 m is preferred to assure sufficient mechanical protection [13].

In general, the equations for systems of ground electrodes are complex and require a numerical solution, as herein explained. However, analytical formulae are available for common configurations, which can be used with the proper precautions by a professional engineer.

To introduce the fundamental concepts, we begin with the study in d.c. of a vertical rod of length L and radius a (with $a << L$) buried into a homogeneous soil with resistivity ρ at a certain depth h, with an injected current I (see Figure 3.2).

The single-layer soil approximation is often unrealistic because homogeneous soils are not common, and multi-layer models may be necessary, including two or more horizontal soil layers. The d.c. analysis is accurate enough up to 50–60 Hz, where the resistance is the most significant component of the grounding system. On the other hand, at higher frequencies, the reactance (inductive

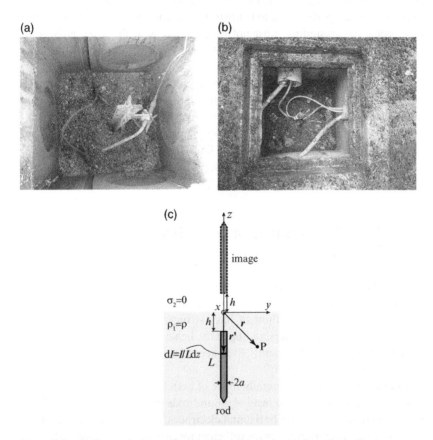

Figure 3.2 (*a*) Copper rod manufactured in solid steel and electrolytically copper-plated, with ends threaded. (*b*) Cross ground rods hot-dip galvanized in profiled steel homogeneous section X, complete with plate welded for the connection with the earth conductor (production in compliance with CEI EN 50164-2 sections 50x50x5). (*c*) Diagram of vertical ground rod with burial depth h.

or capacitive) is usually dominant, and the impedance, rather than the resistance, must be considered to take into account the earth dielectric constant, which is neglected in the d.c. analysis.

In the case of homogeneous soil, the potential dV produced by an elementary current dI must be a solution to the Poisson's equation, and it can be expressed in a cylindrical reference system[2] (Eq. 3.1).

$$dV(\rho,z,z') = \frac{\rho\,dI}{4\pi L}\frac{1}{|\mathbf{r}-\mathbf{r}'|} = \frac{\rho\,dI}{4\pi L}\frac{1}{\sqrt{\rho^2+(z-z')^2}}, \tag{3.1}$$

where $\mathbf{r}=(\rho,\phi,z)$ is the observation point, and $\mathbf{r}'=(\rho',\phi',z')$ is the source point. The presence of the interface between the soil, which is a medium with resistivity $\rho_1=\rho$, and the air, a second medium with infinite resistivity (i.e., $\sigma_2=0$), creates a boundary-value problem, whose solution may be based on the *image theory*.[3] Setting the electrode image as shown in Figure 3.2c, and assuming a uniform current density over the rod, the elementary current is $dI=\frac{I}{L}dz$, and the potential is determined by integrating Eq. (3.1) with respect to z' over the actual rod, from $(-L-h)$ to $-h$ and over its image, from h to $h+L$.

$$V(\rho,z) = \frac{\rho I}{4\pi L}\log\left[\frac{z+h+L+\sqrt{\rho^2+(z+h+L)^2}}{z-h-L+\sqrt{\rho^2+(z-h-L)^2}}\cdot\frac{z-h+\sqrt{\rho^2+(z-h)^2}}{z+h+\sqrt{\rho^2+(z+h)^2}}\right] \tag{3.2}$$

---

2 The cylindrical reference system is characterized by a radial distance ρ, an angular coordinate φ, and a height z.

3 According to the image theory, the ground potential at the point of observation located into the ground is due to a direct contribution form the *source* I_1 locate at $-|z'|$ and a contribution from the *image* I_2, located at $+|z'|$, which takes into account the effect of the finite conductivity of the earth. Both these contributions must be calculated assuming the earth as an infinite and boundless medium, since the effect of the air-soil boundary is taken into account by the image. In the general case, the magnitude of the image contribution is $I_2=KI_z$ where $K=\dfrac{\rho_2-\rho_1}{\rho_2-\rho_1}$.

Assuming $h = 0$, and calculating the potential $V(\rho, z)$ at the top of the ground-rod (i.e., radial coordinate $\rho = a$ and height $z = 0$), the ground resistance R_G is:

$$R_G = \frac{\rho}{4\pi L} \ln\left[\frac{L + \sqrt{a^2 + L^2}}{-L + \sqrt{a^2 + L^2}}\right] \cong \frac{\rho}{4\pi L}\left(2\ln\frac{2L}{a} + \frac{a^2}{2L^2} + O[a^3]\right) \cong \frac{\rho}{2\pi L} \ln\frac{2L}{a} \quad (3.3)$$

The above equation, referred to as the Rudenberg and Datta formula, is affected by a rather large error and should be used with care. It is in fact based on the arbitrary choice of calculating the ground potential V at the head of the rod (*mid-point potential* approximation). Other choices for the locations of the potential along the rod may be found in technical literature, and they produce equally inaccurate results.

The assumption that the current I is uniformly leaked into the ground over the entire length L of the rod is the reason why the results may not be accurate. The leakage current produces equipotential surfaces that are ellipsoids centered on the rod (see Figures 3.3a and 3.3b), whose potentials vary with the burial depth h, as shown in Figure 3.3c. Therefore, the value of R_G depends on where the potential is determined over the rod.

A more accurate method of calculation, which can be used for several shapes of conductors, is the *average potential method* proposed by Dwight. This method consists of calculating the average potential $\bar{V} = \frac{1}{L}\int_{-L}^{0} V(\rho = r_0, z)\,dz$, which

yields the approximate ground-resistance $R_G = \frac{\bar{V}}{I}$, where I is the total current.

Solving the above integral, and approximating the result with the first term of a Taylor expansion, provides the Dwight formula [14] for the ground resistance of the rod:

$$R_G = \frac{\rho}{4\pi L^2}\left[r_0 - \sqrt{4L^2 + a^2} + 2L\ln\left(\frac{2L + \sqrt{4L^2 + a^2}}{r_0}\right)\right] \cong$$

$$\frac{\rho}{2\pi L}[\ln(a) - 1] + \frac{\rho a}{4\pi L^2} - \frac{\rho a^2}{32\pi L^3} + O[a^3] \cong \frac{\rho}{2\pi L}\left[\ln\left(\frac{4L}{a}\right) - 1\right] \quad (3.4)$$

Equation (3.4) is sufficiently accurate for most practical applications.

The ground resistance primarily depends on $\frac{\rho}{L}$ and is relatively independent of the logarithmic component. Consequently, the ground resistance is almost exclusively determined by the length of the rod buried in the ground (i.e., a longer length yields a lower resistance), the resistivity of the ground (i.e., a

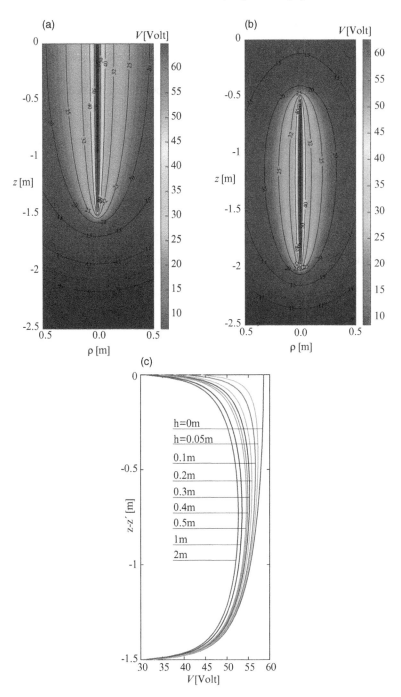

Figure 3.3 Potential distribution of a rod ($L=1,5$cm, $a=1,26$cm, $\rho=100\Omega m$) with burial depth $h=0m$ (a) and $h=0.5m$ (b). Potential profiles over the rod for different burial depths h (c).

lower resistivity yields a lower resistance), and is relatively independent of the radius a of the rod. A reason for using longer rods is the possible presence of deeper layers of soil with higher moisture content, which would decrease the ground resistance of the electrode.

In renewable energy installations, several types of ground electrodes may be found: concrete-encased electrodes (e.g., the supports of the solar racks) or electrodes buried in a volume of soil with more conductive backfill material. In these cases, we can use the formula for the ground-rod resistance proposed by Fagan and Lee [15]:

$$R_G = \frac{1}{2\pi L}\left\{(\rho - \rho_c)\left[\ln\left(\frac{4L}{b}\right) - 1\right] + \rho_c \ln\left(\frac{4L}{a}\right) - 1\right\}, \tag{3.5}$$

where ρ_c and b are, respectively, the resistivity and the radius of the back fill or casing. For example, Eq. (3.5) is applicable to the calculation of the ground resistance of PV mounting system embedded in concrete foundations (Figure 3.4.)

When the ground rod is buried at the depth h, the resistance is given by:

$$R_G = \frac{\rho}{2\pi L}\left\{\ln\left(\frac{4L}{a}\right) - 1 + \ln\left(\frac{h+L}{2h+L}\right) + \frac{h}{L}\ln\left[\frac{4h(h+L)}{(2h+L)^2}\right]\right\}. \tag{3.6}$$

As a ground rod is driven deeper into the earth, its resistance is substantially reduced. The most common ground rod is made of a 1.5 m (5 ft) cylindrical conductor that meets the NEC and IEC codes. A minimum diameter of 1.59 cm (5/8 inch) is required for steel rods, and 1.27 cm (1/2 inch) for copper or copper-clad steel rods.

Other formulae of most practical use are listed in Table 3.1, which can be applied to basic electrodes. In practice, grounding systems usually consist of interconnected conductors ranging in complexity: vertical ground-rods, large ground-grids with ground rods or wells, buried horizontal conductors or rings, and external ground connections. The current leaked by any one buried conductor is the result of the *self ground-potential* and of the *induced ground potential* impressed by all the other conductors forming part of the grounding system. Adding more conductors to a ground system improves its capacity to dissipate current into the soil, and therefore reduces its ground resistance. However, this reduction is limited by the mutual coupling existing among the electrodes, which will cause conductors to leak a current based on their location within the ground system. Thus, the total ground resistance of a complex grounding system is not just the mere parallel of the ground resistances of the

(a)

(b)

(c)

Figure 3.4 (*a*) galvanized steel driven-pile connected to the earth system; (*b*) post in concrete foundation; (*c*) pole ground racking system on concrete block foundation.

Table 3.1 Formulae for the ground resistance of basic horizontal ground electrodes.

1	Circular flat-plate	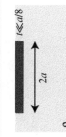	$R_G = \dfrac{\rho}{4a} = \dfrac{\rho}{4}\sqrt{\dfrac{\pi}{A}}$

2 Circular flat-plate deeply buried in soil

$$R_G = \frac{\rho}{8a} + \frac{\rho}{8\pi h}\left(1 - \frac{7}{48}z^2 + \frac{33}{640}z^4 \cdots\right)$$

with $z = \dfrac{a}{h}$

3 Ring-shaped grounding electrode

$$R_G = \frac{\rho}{4\pi^2 b}\ln\frac{32b^2}{ha}$$

4 Buried horizontal wire

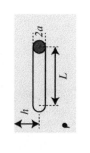

$$R_G = \frac{\rho}{2\pi L}\left[\ln\frac{2L^2}{ah} - 2 + 2w - w^2 + \frac{w^4}{2}\cdots\right]$$
$$\cong \frac{\rho}{\pi L}\left[\ln\frac{\sqrt{2}L}{\sqrt{ah}} - 1\right] \text{ when } h \ll L$$

with $w = \dfrac{h}{L}$

5	Right-Angle Turn of Wire	$R_G = \dfrac{\rho}{4\pi L} \cdot$
		$\left[\ln\dfrac{2L^2}{ah} - 0.2373 + 0.4292w + 0.414w^2 - 0.6784w^4 \ldots\right]$
		with $w = \dfrac{h}{L}$

burial deph h
wire radius a
ground resistivity ρ

6	Three-point star	$R_G = \dfrac{\rho}{6\pi L} \cdot$
		$\left[\ln\dfrac{2L^2}{ah} + 1.071 - 0.418w + 0.952w^2 - 0.864w^4 \ldots\right]$
		with $w = \dfrac{h}{L}$

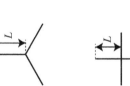

7	Four-point star	$R_G = \dfrac{\rho}{8\pi L} \cdot$
		$\left[\ln\dfrac{2L^2}{ah} + 2.912 - 2.142w + 2.58w^2 - 2.32w^4 \ldots\right]$
		with $w = \dfrac{h}{L}$

8	Six-point star	$R_G = \dfrac{\rho}{12\pi L} \cdot$
		$\left[\ln\dfrac{2L^2}{ah} + 6.851 - 6.256w + 7.032w^2 - 7.84w^4 \ldots\right]$
		with $w = \dfrac{h}{L}$

(continued)

Table 3.1 (Cont.)

9	Eight-point star		$R_G = \dfrac{\rho}{16\pi L} \cdot$ $\left[\ln \dfrac{2L^2}{ah} + 10.98 - 11.02w + 13.04w^2 - 18.72w^4 \cdots \right]$ with $w = \dfrac{h}{L}$
10	N-point star		$R_G = \dfrac{\rho}{N\pi L} \left[\dfrac{1}{2} \ln \dfrac{2L^2}{ah} - 1 + \sum\limits_{m=1}^{N-1} \ln \dfrac{1 + \sin\left(\dfrac{\pi m}{N}\right)}{\sin\left(\dfrac{\pi m}{N}\right)} \right]$
11	Horizontal strip		$R_G = \dfrac{\rho}{16\pi L} \cdot$ $\left[\ln \dfrac{2L^2}{W} + \dfrac{W^2 - \pi Wt}{2(W+t)^2} + \ln\left(\dfrac{1}{z}\right) - 1 + 2z - z^2 + \dfrac{z^4}{2} \cdots \right]$ with $z = \dfrac{h}{L}$
12	Horizontal tabular device		$R_G = \dfrac{\rho}{5.08\pi W} \cdot K_1$ where $K_1 = w \ln\left(\dfrac{1 + \sqrt{1 + w^2}}{w} \right) + \ln\left(1 + \sqrt{1 + w^2}\right) + $ $+ \dfrac{1}{3w} + \dfrac{w^2}{3} - \dfrac{(1 + w^2)\sqrt{1 + w^2}}{3w}$ with $w = \dfrac{W}{L}$

Horizontal strip figure labels: $l \ll W/8$, $W \ll L$, h, L, ρ

Horizontal tabular device figure labels: $l \ll W/8$, W, h, L, ρ

single electrodes, due to the effect of the *mutual resistances*. Mutual resistances introduce a complexity that generally calls for numerical modeling [16, 17].

3.3 The Maxwell Method

A numerical method that is of practical use in the low-frequency range is the *Maxwell Method*, which consists of the solution of the electrostatic voltage integral equation with the *Method of Moments* (MoM) [18–20].

The Maxwell method is based on the general solution of the Poisson's equation for a linear conductor L_c leaking a linear current density J buried in a homogeneous medium with resistivity ρ:

$$V(\mathbf{r}) = \frac{\rho}{4\pi} \int_{L_c} \frac{J}{|\mathbf{r}-\mathbf{r}'|} dl'. \tag{3.7}$$

We assume that the conductor is equipotential at the potential V_G, which allows us to neglect voltage drops in the low-frequency range.

We account for the surface discontinuity between soil and air at $z = 0$ by introducing the image of the source. We obtain:

$$V_G = \frac{\rho}{4\pi} \int_{L_c} G(\mathbf{r},\mathbf{r}') J' dl' = \frac{1}{4\pi} \int_{L_c} R(\mathbf{r},\mathbf{r}') J' dl' \quad \text{with} \quad \mathbf{r} \in L_c, \tag{3.8}$$

where $R = \rho G$ and G is the Green function defined as:

$$G(\mathbf{r},\mathbf{r}') = \frac{1}{\sqrt{(x-x')^2 + (y-y')^2 + (z-z')^2}} + \frac{1}{\sqrt{(x-x')^2 + (y-y')^2 + (z+z')^2}}. \tag{3.9}$$

The integral in Eq. (3.8) is solved with respect to the linear variable l ranging over the entire length of the conductor L_c. To solve Eq. (3.8) with a discrete approach, the grounding system is subdivided into N linear segments (see Figure 3.5a), and every segment can be regarded as a point source leaking an elementary current. The current density $J'(\mathbf{r}')$ is divided in sub-domain basis functions (also known as *expansion function*[4]) $f_n(\mathbf{r}')$ defined over a domain L_n.

---

4 Generally, both entire-domain and sub-domain basis functions can be used to expand the unknown current over the conductor. The choice of basis functions is very wide, ranging from Dirac and pulses (constant) to piecewise linear. All the calculations herein discussed are based on the piecewise linear expansion.

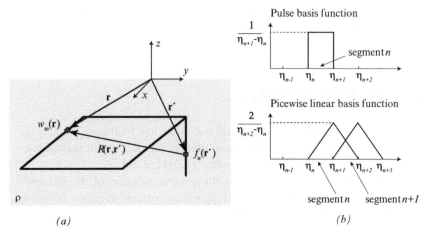

Figure 3.5 General representation of a discretized electrode structure (*a*). Sub-domain basis functions of practical use (*b*).

$$J'(\mathbf{r}') = \sum_{n=1}^{N} \alpha_n f_n(\mathbf{r}') \tag{3.10}$$

where α_n are the weight constants to be calculated.

In many cases, an acceptable choice is to assume the current over each elementary segment to be constant; therefore, pulse basis functions can be defined on each generic *n-th* segment of length L_n (Figure 3.5b):

$$f_n(\mathbf{r}') = \begin{cases} 1 & \text{if } \mathbf{r}' \in \text{segment } n \\ 0 & \text{otherwise} \end{cases} . \tag{3.11}$$

We now introduce a set of N *weighting functions* (or *testing functions*[5]) $w_n(\mathbf{r})$ that can be used to test the integral of Eq. (3.8), through a proper *inner product* (i.e., integration).

$$\underbrace{\int_{L_m} V_G w_m(\mathbf{r})\mathrm{d}l}_{[V_E]_m} = \alpha_n \frac{1}{4\pi} \underbrace{\int_{L_m} w_m(\mathbf{r})\mathrm{d}l \int_{L_n} R(\mathbf{r},\mathbf{r}') f_n(\mathbf{r})\mathrm{d}l'}_{[R]_{mn}} \tag{3.12}$$

---

5 In most of the applications, the Galerkin technique is preferred. It consists of choosing the same testing functions as the basis functions. This applies to both the sub-domain and the entire domain functions.

By testing the integral equation N times with all the testing functions, we obtain the matrix equation.

$$[\mathbf{V}_G] = [\mathbf{R}] \cdot [\alpha] \tag{3.13}$$

The matrix equation can be inverted to obtain the coefficients α_n.

In practice, a unitary potential V_G is assumed and the current density is found through the Eqs. (3.13) and (3.10). The total current I can be then calculated as $I = \sum_{n=1}^{N} \alpha_n |f_n|$, where $|f_n| = \int_{L_n} f_n(\mathbf{r}') d\mathbf{r}'$ is the magnitude of the basis function; the ground-resistance is then obtained as $R_G = {V_G}/{I}$. Once the current coefficients α_n are determined, the ground-potential can be calculated through the general solution of the Poisson's equation.

The numerical method allows a more comprehensive solution since it can be used to validate the results and the accuracy obtained through analytical calculations.

The first approximation that must be discussed is the assumption of the uniform current density over the segments used in the calculations. Figure 3.6 shows the linear current density distribution over a rod for two burial depths in comparison with the uniform approximation. Two effects are visible in the numerical data: (1) the linear current density is not uniform because the parts of the rod deeper into the soil are less affected by the mutual interactions with the other elements, and therefore a higher current can flow into the ground

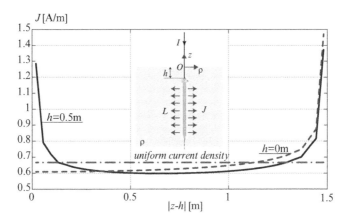

Figure 3.6 Current density distribution over a rod ($L=1.5cm$, $a=1.26cm$, $\rho=100\Omega m$) for two different values of the burial depth h. The uniform current density approximation is reported as well.

(this effect is clearly visible for $h = 0$); (2) the linear current density is higher near the rod tips. From a pure theoretical point of view, under the thin-wire approximation, the leakage current along the rod has a divergent behavior as its distance r from the tips decreases: $\frac{1}{\sqrt[3]{r}}$ [21].

Figure 3.7 shows a comparison of the ground resistance values of a vertical rod obtained with the MOM with those obtained with the analytical calculation per Eq. (3.4), as a function of the length L. Figure 3.8 shows a similar comparison between numerical and analytical results per Eq. (3.6), as a function of the burial depth h.

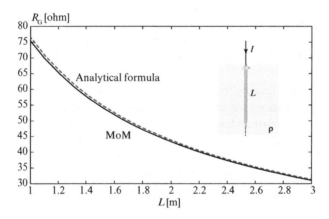

Figure 3.7 Earth resistance of a ground rod ($a=1,26cm$, $h=0m$, $\rho=100\Omega m$) versus length L: comparison between numerical results and calculations per Eq. (3.4).

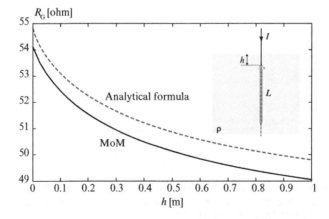

Figure 3.8 Earth resistance of a ground rod ($L=1,5cm$, $a=1,26cm$, $\rho=100\Omega m$) versus burial depth h: comparison between numerical results and calculations per Eq. (3.6).

The analytical approximations provide results with an acceptable accuracy, since the relative error is always lower than 3%, and slightly overestimated ground resistance values yield conservative results. The ground resistance of the rod remarkably decreases with its length and gradually decreases with the burial depth; the shorter is the rod, the higher is the reduction of the resistance with an increase in the burial depth.

Figure 3.9a shows the ground potential distribution $V(r)$ over the soil surface induced by a vertical rod driven at various burial depth with a 1 A

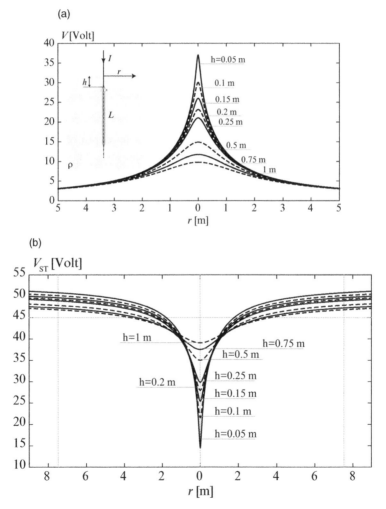

Figure 3.9 Soil ground potential distribution $V(r)$ (a), prospective touch voltage $V_{ST}(r)$ (b) and step voltage (c), for a vertical grounding rod as a function of the distance r from the rod ($L=1.5$m, $a=1.26$cm, $\rho=100\Omega$m, $I=1$A).

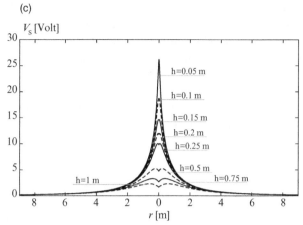

(c)

Figure 3.9 *(Continued)*

leaking current. In general, the larger is the burial depth of the electrode system, the lower are the ground resistance and the surface ground potential. From a safety engineering point of view, it is more effective to determine the *prospective touch voltage* $V_{ST}(r)$, which is defined as the difference between the ground-potential rise V_G of the ground system and the potential $V(r)$ over the soil surface at a distance r from the electrode. Outside the *area of influence* of the ground electrode, approximately at a distance of five times the length of the rod (i.e., 7.5 m in our case), the surface ground potential can be considered negligible, and the prospective touch voltage has approximately reached its asymptotic maximum value. For deeper burial depths, the prospective touch voltage increases near the rod, as the surface potential $V(r)$ decreases faster than the ground-potential rise V_G; far from the rod, instead, $V_{ST}(r)$ decreases, as $V(r)$ is negligible, and the touch voltage is dominated by the reduced V_G with the burial depth, which slowly decreases to its asymptotical value as $h \to \infty$

For a vertical grounding rod, the distribution of the soil surface potential $V(r)$, the prospective touch voltage $V_{ST}(r)$, and the step voltage, as a function of the distance r from the rod, are shown in Figure 3.9. An increase in the ground rod burial depth h may either increase or decrease $V_{ST}(r)$, according to the value of r (i.e., near or far from the rod). The step voltage always decreases with increased burial depths. Step and touch voltages always decrease with increases in soil electrical conductivities and rod lengths.

Numerical computations have been used to provide reference data against which comparing analytical results; numerical computation proves especially useful and effective to design complex ground systems that commonly include buried rings, grids, and multiple rods.

3.4 Multiple Rods: Mutual Resistance

Generally, electrodes in parallel (e.g., ground rods) yield lower ground resistance than a single electrode and may be employed to lower the ground resistance in high-capacity installations. However, two identical rods of length L connected in parallel do not provide a total ground resistance equal to half of that of a single rod, but a higher value; this is true unless the spacing s between the two rods is at least a few rod lengths. This effect is due to the *mutual resistance* that always exists between conductors. In fact, when a buried conductor leaks a current into the soil, the distribution of the current density through its surface is affected by the presence of the other parallel conductors that are also leaking a current. A *shielding effect* among conductors occurs, which disappears only when the conductors no longer interact, that is, when they are ideally buried at an infinite distance from each other. Thus, a correct calculation of the total ground resistance with more conductors in parallel can only be performed by considering the mutual resistances.

No analytical methods for the calculations of the ground resistance of a multi-electrode system exist; however, calculation methods that provide sufficiently accurate results for engineering purposes may be employed.

Mathematically, the potentials of two electrodes in parallel can be expressed as shown in Eq. 3.14.

$$\begin{matrix} V_1 = R_{11}I_1 + R_{12}I_2 \\ V_2 = R_{21}I_1 + R_{22}I_2 \end{matrix} \Rightarrow \begin{bmatrix} V_1 \\ V_2 \end{bmatrix} = \underbrace{\begin{bmatrix} R_{11} & R_{12} \\ R_{21} & R_{22} \end{bmatrix}}_{[R]} \cdot \begin{bmatrix} I_1 \\ I_2 \end{bmatrix}, \tag{3.14}$$

where the resistance matrix **R** is usually symmetrical, that is, $R_{12} = R_{21} = R_m$, since the earth is a reciprocal medium. Introducing the mutual resistance R_m and assuming that the electrodes in parallel are equipotential (i.e., $V_1 = V_2 = V$), the ground resistance of a two-electrode system is given in Eq. 3.15.

$$R_G = \frac{R_{11}R_{22} - R_m^2}{R_{11} + R_{22} - 2R_m}. \tag{3.15}$$

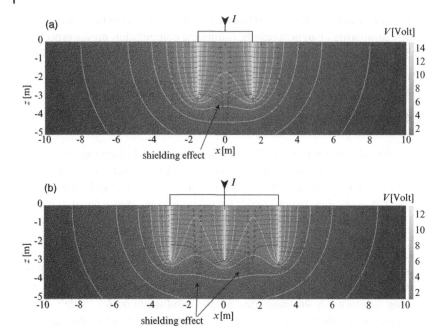

Figure 3.10 Two dimensional map of the potential *V* and the current streamlines produced by two (*a*) and three (*b*) buried rods ($L=3m$, $a=1,26cm$, $s=3m$, $h=0m$, $\rho=100\Omega m$).

It is apparent that the ground system resistance is higher than the pure parallel resistance of the single electrodes; as the spacing *s* between electrodes increases, $R_m \to 0$ and $R_G \to \dfrac{R_{11}R_{22}}{R_{11}+R_{22}}$.

Figure 3.10 shows two-dimensional mapping of the ground-potential and of the current streamlines flowing into the soil for two and three ground-rods in parallel: the shielding effect among electrodes is evident.

In practice, the resistance of two paralleled rods can be calculated with the formulae proposed by Dwight [14] (Eq. 3.16).

$$R_G = \begin{cases} \dfrac{\rho}{4\pi L}\left[\ln\left(\dfrac{4L}{a}\right)+\ln\left(\dfrac{4L}{s}\right)-2+\dfrac{2}{(4w)}-\dfrac{1}{(4w)^2}+\dfrac{1}{2(4w)^4}\right] \cong \\ \qquad\qquad\qquad\qquad \cong \dfrac{\rho}{2\pi L}\left[\ln\left(\dfrac{4L}{\sqrt{as}}\right)-1\right] \qquad \text{for } s<L \\[4mm] \dfrac{\rho}{4\pi L}\left[\ln\left(\dfrac{4L}{a}\right)-1\right]+\dfrac{\rho}{4\pi s}\left(1-\dfrac{w^2}{3}+\dfrac{2}{5}w^4 k\right) \cong \\ \qquad\qquad\qquad\qquad \cong \dfrac{\rho}{4\pi L}\left[\ln\left(\dfrac{4L}{a}\right)-1+\dfrac{L}{s}\right] \qquad \text{for } s>L \end{cases} \qquad (3.16)$$

When the rods are more than two, the formula proposed by Schwarz [22] may be employed (Eq. 3.17).

$$R_{\mathrm{G}} = \frac{1}{N}\frac{\rho}{2\pi L}\left[\ln\left(\frac{4L}{a}\right) - 1\right] + \frac{\rho K_1}{\pi\sqrt{A}}\frac{\left(\sqrt{N} - 1\right)^2}{N}, \tag{3.17}$$

where N is the number of rods in the area A and K_1 is a coefficient developed by Kercel [23] (Eq. 3.18).

$$K_1 = 1.84\sqrt{\frac{ab}{2}}\left[\frac{1}{a}\ln\left(\frac{a + \sqrt{a^2 + b^2}}{b}\right) + \frac{1}{b}\ln\left(\frac{b + \sqrt{a^2 + b^2}}{a}\right) + \frac{a}{3b^2} + \frac{b}{3a^2} - \frac{\left(a^2 + b^2\right)^{\frac{3}{2}}}{3a^2b^2}\right] \tag{3.18}$$

In Eq. (3.18), a and b are respectively the length of the long side and of the short side of the surface identified by the rod configuration.

Figure 3.11 shows the comparison between numerical (MoM) and analytical results of the ground resistance of multiple rods in parallel (i.e., 2, 3, and 4 rods) as a function of s/L, for a given ground-rod length L.

Dwight approximation shows a discontinuous transition between the two formulas when $s \to L$; Schwarz with Kercel approximation is accurate only when the rods are well spaced, that is, $s > 2L$, which seems to be the norm in practical applications. The aforementioned formulae do not consider the burial depth h, which is assumed to be zero. In his original paper [22], Schwarz

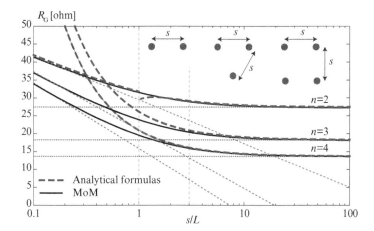

Figure 3.11 Comparison between numerical and analytical results for the ground resistance of multiple rods in parallel vs. s/L. The asymptotical values for small and large spacing s are reported as well ($L=3m$, $a=1,26cm$, $h=0m$, $\rho=100\Omega m$).

proposed the values of the coefficient K_1 also for two burial depths equal to 1/10 and 1/6 of the square root of the area identified by the rod configuration; yet, the coefficients were presented in graphical form and for a limited range of the length-to-width ratio $X = b/a$.

In IEEE-Std 80 [24], the following formulae are included:

$$K_1 = \begin{cases} 1.41 - 0.04X & \text{for } h = 0 \\ 1.20 - 0.05X & \text{for } h = \dfrac{1}{10}\sqrt{A} \\ 1.13 - 0.05X & \text{for } h = \dfrac{1}{6}\sqrt{A} \end{cases} \qquad (3.19)$$

The mutual resistance is a general concept that is not limited to ground-rods, but it also applies to vertical and horizontal electrodes. Figure 3.12 shows the ground resistance of the basic horizontal ground electrodes reported in Table 3.1, for the case of $L = 3\text{m}$. Adding more arms to the star configurations ($N = 2,3,4,6,8$), lowers the ground resistance of the overall system, but the reduction does not scale down as $\dfrac{1}{N}$.

To lower ground resistances, it is a common practice to place rods in a line, a circle, a triangle (arrangement usually named a "triad"), or a square, separated by at least one rod length. Usually, rods in a straight line spaced of at least their length L are more efficient and result in a lower overall ground impedance. Several manuals propose the calculation of the equivalent ground resistance by

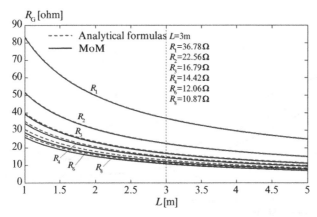

Figure 3.12 Ground resistance of different grounding systems made of bare annealed red copper wire with cross-section of 95 mm² (19 strand wires with diameter of 2.49 mm; overall diameter equal to 12.5 mm): (1) buried horizontal wire; (2) right-angle turn of the wire; (3) three-point star; (4) four-point star; (6) six-point star; (8) eight-point star ($a=1.26cm$, $h=0.5m$, $\rho=100\Omega m$).

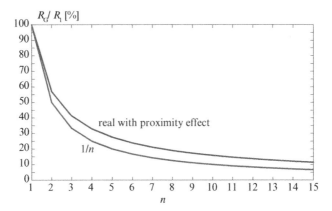

Figure 3.13 Graphs of the ground resistance of a system composed of n rods, one rod length apart in a line, versus n ($L=3m$, $s=L$, $a=1.26cm$, $h=0$, $\rho=100\Omega m$).

multiplying the ground resistance of the single rod by a reduction factor X_R, whose value is usually reported in tabulated or graphical form (Figure 3.13).

Useful formulae for N grounds rods of equal length L, buried at equal space around a circle of diameter D, typical of wind farms, are:

$$R_G = \begin{cases} \dfrac{\rho}{2\pi L}\left\{\ln\left[\dfrac{4L}{\sqrt[N]{Na\left(\dfrac{D}{2}\right)^{N-1}}}\right]-1\right\} & \text{for } D < L \\[3em] \dfrac{\rho}{2N\pi L}\left[\ln\dfrac{4L}{a}-1+\dfrac{L}{D}\sum_{m=1}^{N-1}\dfrac{1}{\sin\left(\dfrac{\pi m}{N}\right)}\right] & \text{for } D > L \end{cases} \qquad (3.20)$$

3.5 Ground Rings and Ground Grid

Grounding systems must be designed to achieve the lowest practical resistance. The way the electrodes are installed and interconnected may have a dramatic effect on the overall performance of the grounding system. According to the NEC (section 250.53), if a single rod, pipe, or plate grounding electrode has a resistance to ground of 25 Ω or less, no supplemental electrode, in parallel, is required. In facilities with sensitive electronics equipment, the ground resistance should not exceed 5 Ω, and to achieve this value, ground rods may not be the best choice.

(a)

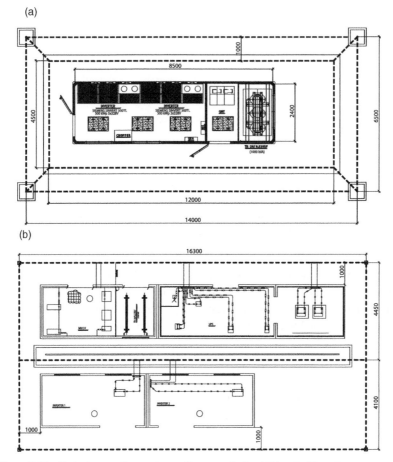

(b)

Figure 3.14 Typical grounding systems of megawatt-sized PV central inverter substations: (*a*) without control room; (*b*) with control room with UPS (dimensions in mm).

In addition, the mitigation of step and touch voltage hazards is accomplished not only through reduction of the ground resistance but also through the proper design and placement of the ground electrodes. For an effective mitigation, a ground ring electrode (sometimes referred to as a *counterpoise*) encircling the structure or equipment may be employed. The purpose of the ground ring is to ensure that an equipotential plane is created for all electrical equipment that are connected to the ground system. The ring electrode equalizes the potential V within the ring by virtue of the principle of electrostatics (only true for *closed surfaces*), which states that if no electrical charges are present within a closed surface S, if the potential V is equal over all points of S, the potential within the surface has also the same value. The ring is however just a rough approximation of a closed surface; thus, the potential equalization is only partial.

(a)   (b)   (c)

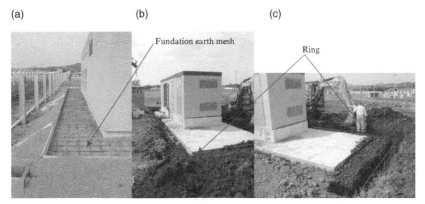

Figure 3.15 Ground systems of megawatt-sized PV central inverter substations: (*a*)-(*b*) Ring ground electrode installed outside the concrete foundation; (*c*) ground electrodes are embedded in concrete foundation.

Ring electrodes are usually placed at a distance of about 1 to 2 m from the structure or equipment (see Figures 3.14, 3.15a, and 3.15b), preferably where water running off the roof can wet the soil.

When the designer needs a better reduction in step and touch voltages, a buried metal grid (or mesh) beneath the structure may provide the optimal ground electrode configuration. A typical arrangement includes both grid and perimeter ring(s), bonded together.

A ground-grid may provide a relatively low ground resistance and minimize potential differences between points, due to the multiple paths between those points. Ground-grids are standard for substations where equalization of voltage differences is essential. Figure 3.16 shows the design of the ground grid of a HV/ MV substation for a wind farm, where the average side of the grid mesh is 5 m. Figure 3.17 shows construction details of the grounding system during its installation. Horizontal electrodes are preferably buried at a depth of 0.5 m to 1 m below ground, so that to have adequate mechanical protection, and it is recommended that ground electrodes be buried below the frost line. Additionally, a layer of high-resistivity material may be spread over the soil surface (Figure 3.16) to introduce a high-resistance contact between the soil and the person's feet, and reduce the body current in case of contact with an energized part.

To minimize the effects of dry or frozen soil on the overall ground resistance, ground rods are also connected to the grid/ring arrangement. Ground rods can, in fact, reach deeper into the earth, where higher soil moisture may be found, or the soil does not freeze. Ground rods, therefore, help maintain a low value of ground resistance during cold seasons.

Rods are installed at the corners of grids/rings, where step and touch voltages usually reach maximum values; ground rods may also be located along

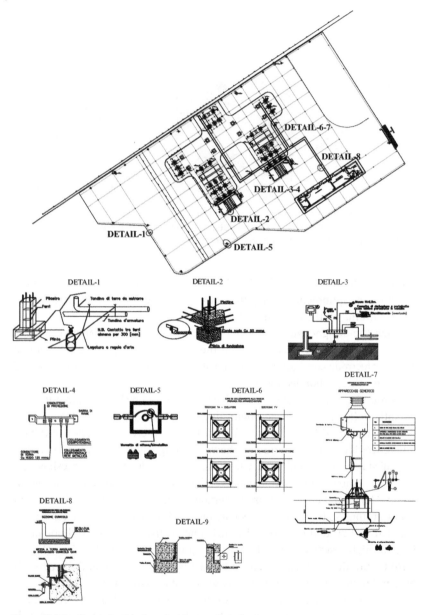

Figure 3.16 Ground-grid of a wind farm substation.

Figure 3.17 Construction details of a grounding system of a wind farm substation.

the perimeter and throughout the grid area. Due to the shielding effect earlier discussed, the current leaked into the soil is larger at the perimeter of the grid and reaches a peak at the corners. The maximum values of touch and step voltages, therefore, occur along the diagonal of grids or rings (Figure 3.18).

The design of meshed ground systems is usually based on the *mesh voltage* V_M, defined as the maximum touch voltage within the mesh (Figure 3.18). Step voltages are inherently less dangerous than mesh voltages, as the current

(a)

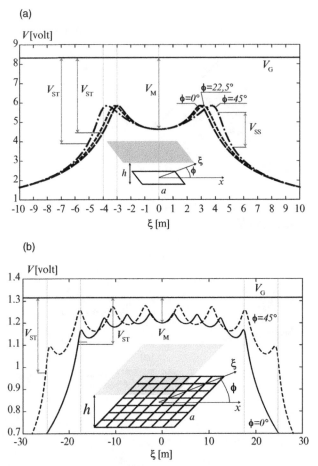

(b)

Figure 3.18 Surface ground potential $V(r)$ for a square ring (a) along the median ($\phi=0°$), the diagonal ($\phi=45°$), and $\phi=22.5°$ (side $a=6$ m, burial depth $h=0.5$ m, $\rho=100\Omega m$, current $I=1A$). The mesh voltage V_M and the prospective touch and step voltages, V_{ST} and V_{SS}, are shown (one meter out of the perimeter). Surface ground potential $V(r)$ for a square ground-grid (b) (mesh side $a=5$ m, number of meshes 7×7, burial depth $h=0.5$ m, $\rho=100\Omega m$, current $I=1A$).

pathway through the human body hardly involves the cardiac region, and therefore they are no longer considered by international standards as a design parameter for ground grids. A review of the fundamental voltage definitions is reported in Figure 3.19.

A ground electrode that may be used in small substations is the reinforcement mesh embedded into the foundation concrete slab. This is an *unintentional* ground electrode, made of metal parts that are not originally intended as

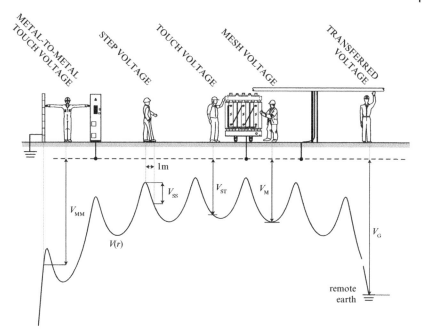

Figure 3.19 Definitions of fundamental voltages.

electrodes, but that, however, are very effective in performing this role. International standards do define the *foundation earth electrode* as a conductive part buried in the soil under a building foundation or, preferably, embedded in concrete of a building foundation, generally in the form of a closed loop, which is maintenance-free and is installed as a part of the building's lifecycle. Foundation earth electrodes must be corrosion-resistant and are effective as the moisture retained by the concrete establishes a conductive connection between the reinforcement bars and the soil. To ensure a permanent conductive connection with the soil, an external ring electrode may be installed outside the concrete foundation and bonded to it.

This is a typical arrangement used in wind turbines' grounding system, where round (Figure 3.20) or rectangular (Figure 3.21) rings are commonly used. For wind turbines, standard IEC 61400-24 recommends that the re-bars in the building foundation be included in the grounding system, because metal elements of large foundation structures greatly contribute to lower the earth resistance of electrodes. This standard recommends an arrangement that comprises either an external ground ring electrode in contact with the soil for at least 80% of its total length or a foundation ground electrode.

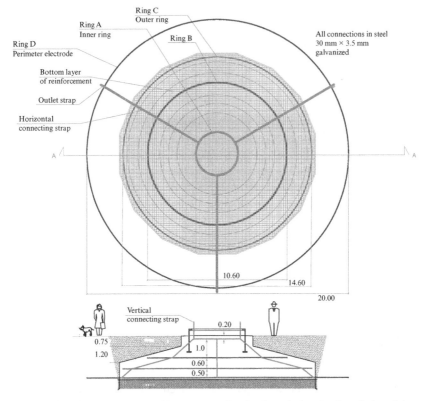

Figure 3.20 Sketch of the earthing-systems for circular wind tower foundation slab.

A round *inner equipotential ring* may be located above the flooring slab of the wind tower foundation at around 15 cm from the wall of the ferrule,[6] fixed to the flooring slab by means of steel clamps bolted to the concrete with poly-amide plugs.

An *outer equipotential ring* may be buried at a distance of 1 m from the outer contour of the tower (burial depth of at least 0.5 m); this burial depth may be increased depending on the meteorological conditions of the site and the soil resistivity.

A *perimeter electrode* may be buried at a distance of 1 m from the edge of the foundation, on the outside, and at a minimum depth of 1 m, which is half a

6 The ferrule connects the tower of the wind generator to its foundation and may be made up of two annular metal flanges, linked by a metal sheet, which forms its wall surface.

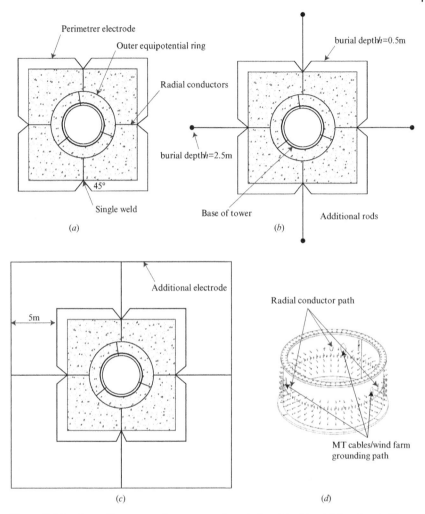

Figure 3.21 Grounding system for a square foundation (a) with additional rods (b) or electrode (c); (d) detail of radial conductor and wind farm grounding path through the ferrule.

meter deeper than the outer equipotential ring (Figure 3.22). A more suitable location of this electrode may be chosen according to the soil properties. However, the minimum grounding electrode configuration consists of two concentric rings: an inner equipotential ring embedded in the concrete and an outer ring embedded in the soil. Ring electrodes and reinforcing bars in the foundation should also be connected to the wind tower metal structure.

(a) (b) (c)

(d) (e) (f)

(g) (h) (i)

Figure 3.22 Views of the construction of a wind tower.

All the ground rings are bonded through multiple *radial conductors* (i.e., at least three or more bare conductors); the electrical connections of the radial conductors to the inner equipotential ring, the outer equipotential ring, and the perimeter electrode are made by aluminothermic welding. The radial conductors must be as short as possible, without any bends or sudden twists in order to minimize their impedance. The radial conductors between inner and outer ground rings are routed through corrugated tubes embedded within the concrete of the foundation; this prevents the galvanic coupling with the steel of the re-bars and facilitates their installation or replacement after pouring the concrete. The number and arrangement of the connections between outer equipotential ring and perimeter electrode primarily depend on the geometry of the foundation: (i) for circular foundations, three radial conductors are normally used, in a wye configuration; (ii) for square and octagonal foundations, four radial conductors are generally used. In the latter case, the conductors must be perpendicular to the mid-point of one of the sides of the electrode perimeter, and should not be connected to the any of the corners. This is to ensure the minimum conductor length and facilitate the execution of the welding work.

Rings are usually made of a bare-stranded copper conductor with a minimum cross-sectional area of 50 mm^2 (recommended 70 mm^2), with each strand of 1.7 mm diameter. Alternatively, other materials for the ring may be used in lieu of copper (e.g., hot-dip galvanized steel strip), as long as size and arrangement follow technical standards. All the electrical connections among the rings and with additional electrodes are made using aluminothermic welding. Clamps should not be used due to their high impedance at high frequencies and reduced thermal capability: welded connections can withstand up to 700°C, while bolted joints may fail when they reach a temperature of 250°C.

As a general rule, a ground resistance not exceeding 10 Ω should be achieved, which would also meet the lightning protection requirements. Whenever the outer ground ring is not sufficient for the above goal, for example, due to high soil resistivity, additional copper electrodes should be incorporated in the design of the grounding system. Based on the soil and type (e.g., rocky, sandy) and soil resistivity, the design of the grounding system may include ground rods connected to the perimeter electrode, or additional concentric rings.

Engineers should ensure that a ground resistance ≤10 Ω is achieved before connecting the wind turbine generator to the electrical system.

The interconnection of the grounding systems of turbines in a wind farm (Figure 3.23) is usually made with copper or aluminum bare conductors, buried at approximately 1 m; this is, however, an expensive configuration.

The ground resistance R_G of a ground grid primarily depends on the surface A of its enclosed area, which is usually known in the early stage of the design. An approximated value can be estimated by using the expression of the ground resistance of a circular metal plate at zero burial depth,

$$R_G = \frac{\rho}{4}\sqrt{\frac{\pi}{A}}. \tag{3.21}$$

The above formula has been corrected by Laurent and Nieman, who added a second term that depends on the total buried length of the conductors L_C forming part of the ground grid

$$R_G = \frac{\rho}{4}\sqrt{\frac{\pi}{A}} + \frac{\rho}{L_C}. \tag{3.22}$$

The approximation introduced by the above expression is usually conservative and has been expanded by Sverak, who also included the effect of the grid depth h,

$$R_G = \rho\left[\frac{1}{L_C} + \frac{1}{\sqrt{20A}}\left(1 + \frac{1}{1 + h\sqrt{\frac{20}{A}}}\right)\right]. \tag{3.23}$$

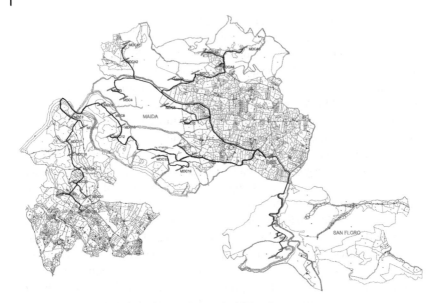

Figure 3.23 General view of an onshore wind farm site.

The ground resistance of the grid can be calculated with the formula proposed by Schwarz, which proved to be the most accurate,

$$R_G = \frac{\rho}{\pi L_C} \left[\ln\left(\frac{2L_C}{a'}\right) + \frac{K_1 L_C}{\sqrt{A}} - K_2 \right]. \tag{3.24}$$

In Eq. (3.24), a' is the radius of the wire, if the burial depth $h = 0$, otherwise $a' = \sqrt{2ha}$, where a is the radius of the wire; K_1 is given by Eq. (3.18) or Eq. (3.19), and K_2 is given be the formula proposed by Kercel (3.25).

$$K_2 = \ln\left[\frac{4(a+b)}{b}\right] + 2K_1 \frac{a+b}{\sqrt{A}} - \ln\left[\frac{a + \sqrt{a^2 + \left(\frac{b}{2}\right)^2}}{\frac{b}{2}}\right] - \frac{1}{2}\ln\left[\frac{\frac{b}{2} + \sqrt{a^2 + \left(\frac{b}{2}\right)^2}}{-\frac{b}{2} + \sqrt{a^2 + \left(\frac{b}{2}\right)^2}}\right] \tag{3.25}$$

In IEEE-Std 80 [24], the following formulae are reported as well, in terms of the aspect ratio X:

$$K_2 = \begin{cases} 5.50 + 0.15X & \text{for } h = 0 \\ 4.68 + 0.10X & \text{for } h = \frac{1}{10}\sqrt{A}. \\ 4.40 - 0.05X & \text{for } h = \frac{1}{6}\sqrt{A} \end{cases} \tag{3.26}$$

Schwarz equation provides very accurate results for the ground resistance, whereas the other formulae are less accurate.

For example, Figure 3.24 shows the comparison among the proposed analytical formulae for a rectangular ring buried at $h = 0.5\ m$ below grade with a fixed shorter side a, as a function of the longer side b, and the numerical solution; the graph shows that Schwarz equation (3.24) provides the most accurate estimate.

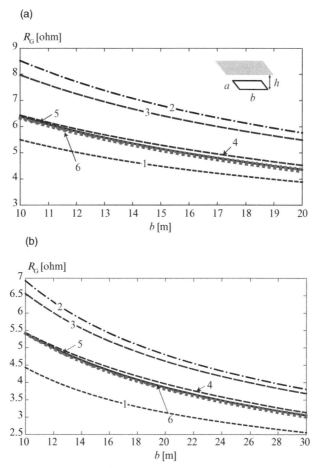

Figure 3.24 Ground resistance of a rectangular ring of shorter side a, longer variable side b, and burial depth h. Comparison among numerical (solid line) solution and different analytical formulae: (1) circular metal approximation in Eq. (3.21); (2) Laurent and Nieman formula in Eq. (3.22); (3) Sverak formula in Eq. (3.23); (4) Schwartz approximation in Eq. (3.24) with Kercel coefficients in Eqs. (3.18) and (3.25); 5) Schwartz approximation in Eq. (3.24) with Schwartz coefficients for $h=0$; 6)

Schwartz approximation in Eq. (3.24) with Schwartz coefficients for $h = \dfrac{1}{10\sqrt{A}}$.

Parameters: bare annealed red copper wire with cross section 95 mm², burial depth $h=0.5$ m, $\rho=100\Omega m$, short side a equal to 6.5 m (a) and 10 m (b).

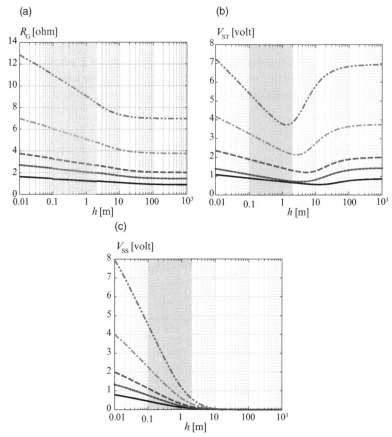

Figure 3.25 Ground resistance (a), prospective touch voltage (b) and prospective step voltage (c) of a rectangular ring ($a=b=5,10,20,30,50$ m, $\rho=100\Omega m$) for different burial depth h (depths between $0.1 \div 2$ m are highlighted).

Deeper burial depths do lower both the ground resistance and the prospective step voltages, and may decrease the prospective touch voltages, depending on the burial depth (Figure 3.25.)

3.6 Complex Arrangements: Rings and Ground Grids Combined with Rods and Horizontal Electrodes

When rods, of average length L, are present along the perimeter of the grid or throughout the grid area, the Schwarz expression of the *mutual resistance* R_{Gm} between the grid and the rods should be considered:

$$R_{Gm} = \frac{\rho}{\pi L_C}\left[\ln\left(\frac{2L_C}{L}\right) + \frac{K_1 L_C}{\sqrt{A}} - K_2 + 1\right] \qquad (3.27)$$

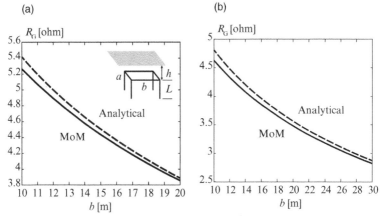

Figure 3.26 Ground resistance of a rectangular ring of shorter side a, longer variable side b, and burial depth h with, four $L=3$ m long rods buried at the corners (parameter as in Figure 3.24): comparison between numerical results and Schwarz equation predictions.

The combined ground resistance of the grid and the rods is given by Eq. (3.15), where R_{11} is the mesh resistance in Eq. (3.24), R_{22} is the resistance of the rod in Eq. (3.17), and R_{12} is the mutual resistance in Eq. (3.27). The coefficients K_1 and K_2 have been introduced, respectively, in Eqs. (3.18)–(3.19) and (3.25)–(3.26). The value of the total ground resistance is lower than that of the ground resistance of either component alone but still higher than that of their parallel combination. The accuracy of the Schwarz equations is excellent (Figure 3.26.)

A low ground resistance is not per se a guarantee of safety. In order to ensure personnel safety, it is essential to keep touch and step potentials within their permissible safe values. Only in the cases when the ground potential rise is less than the permissible touch voltage, no further analysis or design of the grounding system is necessary. The mesh within a ground grid is usually where the worst possible touch voltage occurs (excluding the case of transferred potentials outside the grid area). Therefore, the mesh voltage is normally used as the basis of the design, as indicated in international standards. For equal-mesh ground grids, the mesh voltage V_M increases along the meshes from the center to the corner of the grid, and it is usually equal to the potential difference between the grid conductor and the mesh center.

$$V_M \doteq \rho \frac{K_m K_i}{L_M} I_g, \qquad (3.28)$$

where,

$$K_m = \frac{1}{2\pi} \left\{ \ln \left[\frac{D^2}{16hd} + \frac{(D+2h)^2}{8Dd} - \frac{h}{4d} \right] + \frac{K_{ii}}{\sqrt{1+h}} \ln \left[\frac{8}{\pi(2n-1)} \right] \right\} \text{ and } \quad (3.29)$$

$$K_i = 0.644 + 0.148n, \tag{3.30}$$

where D is the spacing between parallel conductors, d is the diameter of the grid conductor, h is the burial depth, and L_M is the effective buried length. In Eq. (3.29), $K_{ii} = 1$ for ground grids with ground rods along the perimeter or at the grid corners, or both along the perimeter and throughout the grid area; otherwise, $K_{ii} = (2n)^{\frac{-2}{n}}$. n is the effective number of parallel conductors in a given ground grid; if the ground grid has an irregular shape, n represents the number of parallel conductors of an equivalent rectangular ground grid.
n can be calculated as:

$$n = \underbrace{\frac{2L_C}{L_p}}_{n_a} \underbrace{\sqrt{\frac{L_p}{4\sqrt{A}}}}_{n_b} \underbrace{\left[\frac{L_x L_y}{A}\right]^{\frac{0.7A}{L_x L_y}}}_{n_c} \underbrace{\frac{D_{max}}{\sqrt{L_x^2 + L_y^2}}}_{n_d}, \tag{3.31}$$

where L_C is the total length of the horizontal conductors of the grid; L_p is the peripheral length of the grid, A is the area of the grid; L_x and L_y are the maximum length of the grid in the x and y direction, respectively; D_{max} is the maximum distance between any two points on the grid.
The effective length L_M for mesh voltage in Eq. (3.28) is:

$$L_M = L_C + L_R, \tag{3.32}$$

where L_R is the summation of the lengths of all the ground rods.

Equation (3.32) holds true when there are no ground rods, or only a few ground rods scattered throughout the grid, but none located at the corners or along the perimeter of the grid; otherwise, the effective length L_M is:

$$L_M = L_C + \left(1.55 + 1.22\frac{L}{\sqrt{L_x^2 + L_y^2}}\right)L_R. \tag{3.33}$$

From a practical standpoint, it is of interest to evaluate how additional rods may affect the performance of the grounding system. Figure 3.27 shows the prospective touch and step potentials along the diagonal of three different grounding systems: (i) four rods placed at the vertices of a rectangle, (ii) a rectangular ring, and (iii) the combination of ring and rods. Besides the obvious reduction in the ground resistance R_G (i.e., 9.8 Ω, 6.3 Ω and 5.2 Ω, respectively), we observe that the rectangular ring performs better in terms of touch and step voltages compared to the four-rods configuration, and that the combination of ring and ground rods is the most efficient design of the grounding system.

(a)

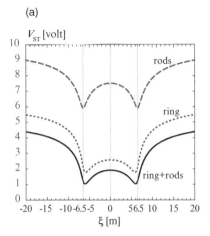

(b)

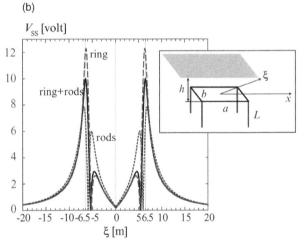

Figure 3.27 Prospective touch V_{ST} (a) and step V_{SS} (b) voltages along the diagonal of three grounding systems: rectangular ring, four rods buried at the corners, ring plus rods ($a=5$ m, $b=12$ m, $h=0.5$ m, $L=3$ m, $\rho=100\Omega m$).

Formula (3.27) is applicable to several configurations often encountered in practical design of grounding systems. The mutual resistance occurring between a buried ring of diameter $2b$ (see Table 3.1, Eq. 3) combined with vertical rods of length L equally spaced (Eq. (3.20) with $D > L$) is:

$$R_{Gm} = \frac{\rho}{4\pi^2 b} \ln\left(\frac{32eb^2}{Lh}\right),$$ (3.34)

where e is the Euler's number.

The above formula is also applicable to a ring connected to N buried radial electrodes of equal length L radiating horizontally and symmetrically from a common point (Table 3.1, Eqs. 6–10).

The mutual resistance of a buried horizontal wire (Table 3.1, Eq. 4) connected to N vertical rods (Eq. (3.17)) is:

$$R_{\mathrm{Gm}} = \frac{\rho}{\pi L_e} \ln \left(\frac{2L_e}{\sqrt{2\frac{L}{e}h}} - 1 \right),$$ (3.35)

where L_e is the horizontal electrode length and L is the length of the rod.

If the N rods are equally spaced along the horizontal electrode with spacing s, (3.17) can be written as:

$$R_{\mathrm{G}} = \frac{\frac{\rho}{2\pi L}\left[\ln\left(\frac{4L}{a}\right) - 1\right]}{N} + \frac{\rho}{N\pi s}\sum_{m=2}^{N}\frac{1}{m}.$$ (3.36)

References

1 IEEE Emerald Book. IEEE recommended practice for powering and grounding electronic equipment – Redline. *IEEE Std 1100-2005 (Revision IEEE Std 1100-1999) – Redline*, 1–703, 2006.

2 Hernández, J.C. and Vidal, P.G. (2009). Guidelines for protection against electric shock in PV generators. *IEEE Transactions on Energy Conversion* 24 (1): 274–282.

3 Zipse, D.W. (November 2003). Earthing or grounding? *IEEE Industry Applications Magazine* 9 (6): 57–69.

4 Rappaport, E. (November 2015). Myths of equipment grounding and bonding. *IEEE Transactions on Industry Applications* 51 (6): 5212–5217.

5 Mitolo, M. (2008). Protective bonding conductors: An IEC point of view. *IEEE Transactions on Industry Applications* 44 (5): 1317–1321.

6 Armstrong, K. (2000). Installation cabling and earthing techniques for EMC. *IEE Seminar on Shielding and Grounding (Ref. No. 2000/016)*, 4/1–410.

7 Enrique, E.H. and Walsh, J.D. (2015). Analysis of touch potentials in solar farms. *IEEE Transactions on Industry Applications* 51 (5): 4291–4296.

8 Kontargyri, V.T., Gonos, I.F., and Stathopulos, I.A. (November 2015). Study on wind farm grounding system. *IEEE Transactions on Industry Applications* 51 (6): 4969–4977.

9 Hoerauf, R. (2014). Considerations in wind farm grounding designs. *IEEE Transactions on Industry Applications* 50 (2): 1348–1355.

10 Kondylis, G.P., Damianaki, K.D., Androvitsaneas, V.P., and Gonos, I.F. (2018). Simplified formulae method for estimating wind turbine generators ground resistance. *IEEE Transactions on Power Delivery* 33 (6): 2829–2836.

11 White, J.R. and Jamil, S. (2016). Do's and don'ts of personal protective grounding. *IEEE Transactions on Industry Applications* 52 (1): 677–683.

12 Zipse, D.W. (2002). Earthing – grounding methods: A primer. *IEEE Technical Conference Industrial and Commerical Power Systems*, 158–177.

13 Dwight, H.B. (1936). Calculation of resistances to ground. *Transactions of the American Institute of Electrical Engineers* 55 (12): 1319–1328.

14 Fagan, E.J. and Lee, R.H. (1970). The use of concrete-enclosed reinforcing rods as grounding electrodes. *IEEE Transactions on Industry and General Applications* IGA-6 (4): 337–348.

15 Velazquez, R., Reynolds, P.H., and Mukhedkar, D. (1983). Eearth-return mutual coupling effects in ground resistance measurements of extended grids. *IEEE Transactions on Power Apparatus and Systems* PAS-102 (6): 1850–1857.

16 Chow, Y.L., Elsherbiny, M.M., and Salama, M.M.A. (November 1995). Efficient computation of rodbed grounding resistance in a homogeneous earth by Galerkin's moment method. *IEE Proceedings – Generation Transmission and Distribution* 142 (6): 653–660.

17 Ney, M.M. (1985). Method of moments as applied to electromagnetic problems. *IEEE Transactions on Microwave Theory and Techniques* 33 (10): 972–980. doi:10.1109/TMTT.1985.1133158.

18 Jin, J. (2010). The method of moments. *Theory and Computation of Electromagnetic Fields*, IEEE, 399–462.

19 Harrington, R. (1990). Origin and development of the method of moments for field computation. *IEEE Antennas and Propagation Magazine* 32 (3): 31–35.

20 Meixner, J. (July 1972). The behavior of electromagnetic fields at edges. *IEEE Transactions on Antennas and Propagation* 20 (4): 442–446.

21 Schwarz, S.J. (1954). Analytical expressions for the resistance of grounding systems. *Transactions of the American Institute of Electrical Engineers Part III Power Apparatus and Systems* 73 (2): 1011–1016.

22 Kercel, S.W. (1981). Design of switchyard grounding systems using multiple grids. *IEEE Transactions on Power Apparatus and Systems* PAS-100 (3): 1341–1350.

23 IEEE Power and Energy Society. (2013). IEEE guide for safety in AC substation grounding. *IEEE Std 80-2013 (Revision IEEE Std 80-2000/Inc. IEEE Std 80-2013/Cor 1-2015)*, February, 1–226.

4

Lightning Protection Systems

CONTENTS

When from thy shore the tempest beat us back,
I stood upon the hatches in the storm.

Henry VI, Shakespeare

Electrical Safety Engineering of Renewable Energy Systems. First Edition. Rodolfo Araneo and Massimo Mitolo.
© 2022 by The Institute of Electrical and Electronics Engineers, Inc. Published 2022 by John Wiley & Sons, Inc.

4.1 Review of Natural Lightning Physics, Modeling and Protection

For a better understanding of the next sections, we briefly review and discuss the basic lightning physics and the related terminology used for engineering applications. A large amount of literature is nowadays available on this topic, which provides an in-depth description of all relevant aspects of lightning, such as lightning discharge physics, lightning occurrence characteristics, lightning statistical parameters, *lightning electromagnetic pulse* (LEMP) and induced effects, protection against LEMP [1, 2], etc. The research on lightning can be dated back to the first studies of K. Berger, R. B. Anderson, H. Kroninger, and M. A. Uman in the 1970s, and earlier, when the first observations were based on flash counter records. Additionally, today, several major books on lightning and its effects can be used as a reference [3–9].

Lightning is a transient high-current electric discharge whose path length is measured in kilometers. The *cloud-to-ground* (CG) lightning (see Figure 4.1) is the most relevant type of lightning for engineering applications since it may endanger human life and affect lightning protection systems on earth surface. When the discharge occurs within the same cloud, the lightning is rereferred to as *intra-cloud* discharge, whereas that that involves two or more clouds is called *inter-clouds* or *cloud-to-cloud* discharge. Cloud discharges account for about 75% of global lightning occurrences and do not involve the earth surface [10].

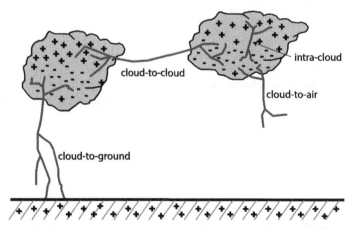

Figure 4.1 Types of lightning flashes comprising (*a*) cloud-to-ground, (*b*) cloud-to-cloud, (*c*) intra-cloud, and (*d*) cloud-to-air.

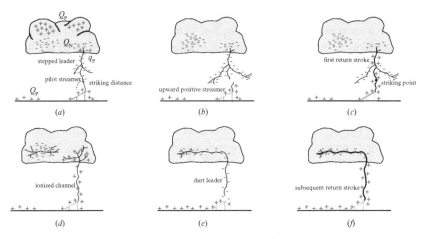

Figure 4.2 Evolution of cloud-to-ground lightning with schematic representation of various stages.

The lightning mechanism is understood to be due to bipolar charged thunderstorm clouds (i.e., cumulonimbus), which induce an opposite charge on the ground. The earth is normally negatively charged with respect to the atmosphere, and so is the bottom layer of clouds. During thunderstorms, the negative charges in the cloud attract positive charges, which will concentrate on vertical objects on the ground. In the idealized model of the cloud charges shown in Figure 4.2, the main charge regions Q_P and Q_N are of the order of many tens of coulombs of positive and negative charge, respectively, and q_P is a smaller positive charge. After a gradient of approximately 10 kV/cm is set up in the cloud, the surrounding air becomes ionized, and when the local electric field reaches a critical value of about 30 kV/cm, an electric discharge is initiated. It is generally assumed that a preliminary breakdown takes place in the lower part of the cloud between the charge centers Q_N and q_P within the cloud. At this point, a *pilot streamer* starts from the cloud, and it moves toward the earth in the air (see Figure 4.2a). This phenomenon cannot be detected with the naked eye, due to its very low luminosity; however, a travelling spot may be noticed. The current in the streamer is of few amperes, and its speed is about 1.6×10^5 m/s. The pilot streamer is followed by a discharge path made of a series of short luminous steps, the so-called *downward stepped leader*. Depending upon the state of ionization of the air surrounding the streamer, the stepped-leader channel branches in a downward direction during its development toward ground. The stepped leader moves rapidly downward with an average velocity of about 5×10^5 m/s and a current of around 100 A, in steps of 50 m to 100 m, and pauses after each step for a few tens of microseconds. For a stepped leader from a cloud of 3 km above ground, about 60 μs may be needed to reach the ground. Since the leader tip is in excess of 10 MV with respect to the earth, as it approaches ground, it raises the electric field strength at the surface of the earth. At a critical point in the development of a downward leader, a responding *upward positive streamer* (see Figure 4.2b) may be lunched from grounded structures (i.e., the *attachment process*). This process will determine the location of the strike, unless another competing upward streamer comes first and makes the first connection to an adjacent structure. The point where the lightning flash touches (i.e., attaches to) the structure is called *attachment point.*

The upward streamers travel toward the downward propagating leader at a speed ranging between 20 and 60×10^3 m/s. The distance between the tip of the downward leader and the origin of the successful upward streamer is defined as the *striking distance.* Its calculation allows the assessment of the probability of a flash to a structure and the evaluation of the efficacy of the protection afforded by *lightning protection systems* (LPS), e.g., grounded air terminals, overhead ground wires, or a Faraday cage.

When the downward stepped leader and the upward positive streamer meet, the two channels join at the *striking point* (see Figure 4.2c), and this initiates the *return stroke*, a fast-transient current that progresses upward from ground to cloud, similarly to a traveling wave on a transmission line. The return stroke travels through the established ionized channel at speed ranging from 0.05 to 0.5 times the speed of light c_0 in vacuum. The return stroke flows from ground to neutralize the charge associated with the negative descending leader, which results in the collapse of the electric field. The current in the return stroke is in the order of a few kA to 300 kA, and the temperatures within the channel range between 15,000°C to 20,000°C. The high current and temperature are responsible for the destructive effects of lightning, causing explosive air expansions and intense luminosity.

It must be noted that even though the properties of the leader channel are used to model the lightning strike process to ground structures, the consequential damages are strongly current-dependent through the *action-integral* $\int i_0 dt$ of the return stroke.

The process so far described is the *downward negative lightning discharge* (Figure 4.2a), which is believed to be the most common, accounting for about 90% of the cloud-to-ground discharges, globally. Less than 10% are *downward positive lightning*. The origin of downward positive lightning is not yet fully understood; this phenomenon can be dominant during the cold season and the dissipating stage of a thunderstorm. It is believed that positive cloud-to-ground discharges are associated with the aftereffects of prolonged cloud-to-cloud discharges, which in the end result in a downward positive discharge. They are of interest because they are associated with the highest recorded lightning currents (near 300 kA), and their current waveform is used by the Std. IEC 62305-1 to define the *lightning protection level* (LPL) I and by the Std. IEC 61643-11, which defines the Class I *surge protective devices* (SPD).

Lightning can also be initiated at the ground, usually from tall objects of height of about 100 m, or from objects of moderate height located on mountain tops, by upward-going stepped leaders that can be either positively or negatively charged (*upward lightnings*). This is in line with the saying that "*lightning doesn't not strike tall objects*," but instead tall objects induce lightning conduction path toward themselves. The upward stepped leader can be initiated by a slow-increasing electric field produced by the thundercloud charge (*self-initiated upward flash*) or can occur under the fast-electric field change caused by nearby lightning discharges (*other-triggered upward flash*).

The classification of lightnings is done according to the direction of the initial leader discharge in the attachment process and the polarity of the charge transferred from the thundercloud to the ground; a schematic representation of types of lightning flashes is reported in Figure 4.3. Observations showed that negative upward flashes are more common than positive ones [11, 12].

It is widely reported that wind turbines (WTs) trigger upward lightning flashes, especially under the winter thunderclouds [13–16]; it should be kept in mind that an 8 MW-wind turbine may have a height of 220 m and a rotor diameter of 164 m. Upward lightning flashes are considered to be the main source of lightning damages to modern WTs [12], mainly because the height of the tip of the blade may be much larger than any surrounding structure. The incidence of upward lightning significantly increases with the increase in both the height of the hub and the length of the blades [17]; in addition, the rotation of blades facilitates the initiation of upward lightning [18]. An important role is

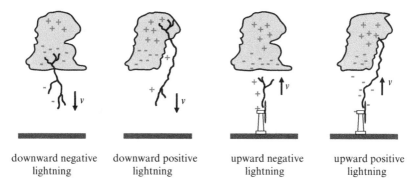

downward negative   downward positive   upward negative   upward positive
lightning   lightning   lightning   lightning

Figure 4.3 Types of lightning flashes.

also played by the space charge produced by the wind tower during thunderstorms due to the corona effect. In static conditions, this charge is distributed in layers that shield the local electric field at the lightning receptors incorporated into the blades and weakens the initiation of streamers; however, the blade rotation moves away from the receptors this space charge, and the shield effect decreases. Thus, it is believed that upward lightning is easier to be triggered from operating WTs than from static objects. Multiple upward leaders are observed to be simultaneously triggered from different WTs under the influence of thunderstorms [19].

After the first discharge process (see Figure 4.2d), there may be charges in the cloud near the first neutralized charged cell that will be further neutralized thanks to the ionized path. Consequently, after a delay of about 50 ms, further subsequent strokes at 10 to 300 μs intervals may occur along the same path for most of its length. The leader of the subsequent strokes is known as the *dart leader* because of its dart-like appearance (see Figure 4.2e), and the subsequent discharge (or return stroke) starts when the dart leader meets the positive streamer that is rising up from ground (see Figure 4.2f). The dart leader follows mainly the path of the first stepped leader with a speed about ten times faster than the stepped leader. The corresponding path is usually not branched and is brightly illuminated. On average, about 80% or more of negative cloud-to-ground lightning flashes may contain from three to five subsequent return strokes, while positive downward flashes typically contain no subsequent strokes (an example is provided in Figure 4.4).

In the case of negative subsequent strokes, two more possible modes of charge transfer to ground may occur, in addition to the dart leader/return stroke sequences: the *continuing currents*, and the *M-components*. The lightning continuing current is a quasi-stationary arc between the cloud charge

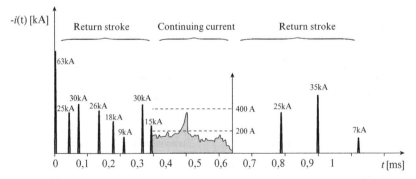

Figure 4.4 Current of a multiple stroke negative downward lightning. It is composed of 11 return strokes and one continuing current.

source region and the ground; M-components are spikes or surges that are created in the continuing current and transport negative electric charge from the cloud to ground. The M-components are likely to occur during the *initial continuous-current* stage of the upward negative lightning. When the upward positive leader bridges the gap between the wind tower and the negative charge source in the bottom of the cloud, an initial continuous current is established, that typically lasts for several tens to several hundreds of milliseconds with an average amplitude of several hundred of amperes. The initial continuous current usually includes superimposed impulsive processes (the α-components) that resemble the M processes. The initial continuous current, which constitutes the initial stage of the upward flash, is usually followed, after a no-current interval, by one or more downward-leader–upward-return-stroke sequences (the β-components).

Subsequent strokes are often referred to as *hot lightning strokes* (as opposed to the first return stroke defined as the *cold lighting stroke*) because, even though the current is relatively smaller, they last for a few milliseconds, and therefore carry a relatively larger energy, and have greater maximum current derivatives $\dfrac{di}{dt}\bigg|_{max}$; cold lighting strokes usually cause higher induced voltages that may result in lightning-related insulation failures. In fact, although the peak current of the first strokes has a median value that is about 3 times greater than that of the subsequent strokes, the latter have median front-times 5 to 8 times shorter.

Traditional lightning parameters needed in engineering applications to assess the lightning threat to a structure are: current peak i_{max}, charge $Q = \int i(t)\,dt$, specific energy $W/R = \int i^2(t)\,dt$, and maximum current derivative $\dfrac{di}{dt}\bigg|_{max}$. Generally, the peak current i_{max} is important for the design of the grounding system. The lightning current carried by a down-conductor is injected into the soil thanks to the grounding system, and causes a transient potential rise across the ground impedance; this may lead to side flashes, especially when metalwork enter a structure and is not equipotentially bonded to the grounding system; side flashes may damage electric or electronic systems, and cause fires. The peak current determines the maximum value of this common mode (or *direct-axis*) potential rise.

The charge Q is responsible for the melting effects at the impact points of the lightning channel. The magnitude of the input energy of the arc root is given by the product of the voltage drop across anode and cathode and the charge Q.

The specific energy W/R causes mechanical stress and heating effects across conductors, when the lightning current flows through them. The maximum current derivative (or front steepness) determines the strength of such effects, i.e., the maximum value of the induced voltage into loops.

Mathematical models have been put forward to account for the return stroke. In lightning studies, strokes to ground are of importance due to their direct and induced effects. The engineering model of lightning return stroke assumes that the lightning channel is straight, vertical, and normal to the ground plane. The return stroke current $i(z,t)$ is assumed to be a function of vertical coordinate z and time t and is related to the current $i(0,t)$ generated at the striking point at ground level. The relationship of $i(z,t)$ to $i(0,t)$ may be described based on a modified transmission line model, with exponential current decay

$i(z,t) = e^{-\frac{z}{\lambda}} i\left(0, z - \frac{t}{v}\right) u_0\left(z - \frac{t}{v}\right)$, where λ is the attenuation height (in the range

of $1 \div 2$ km), v is the speed of the return stroke (around 1.3×10^8 m/s) and u_0 is the Heaviside step function. A review of the more general traveling current source model is reported in [20].

In lightning protection studies, different mathematical expressions have been used to describe the return stroke current waveform $i(0,t)$ (e.g., CIGRE 33-01, IEC Std. 62305-1). The lightning waveform is usually defined by the front duration t_f (also referred to as T_1), expressed as a function of the 10–90% rise time and pulse length t_p (time to 50% of the maximum value), also referred to as *tail duration* T_2), and peak value I_{peak}.

The standard lightning waveform is shown in Figure 4.5. Although lightning waveforms may differ from actual transients, the standardized forms (e.g., in Stds. EN 61643, EN 62305) are based upon years of observations and measurements and generally provide a fair approximation of the actual transient. Both the IEC and the IEEE have standardized on the 10/350 µs waveform and the

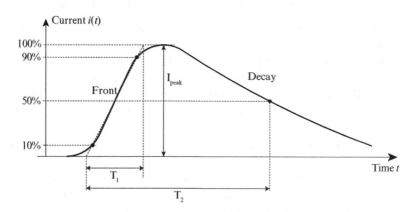

Figure 4.5 Short return-stroke current waveform.

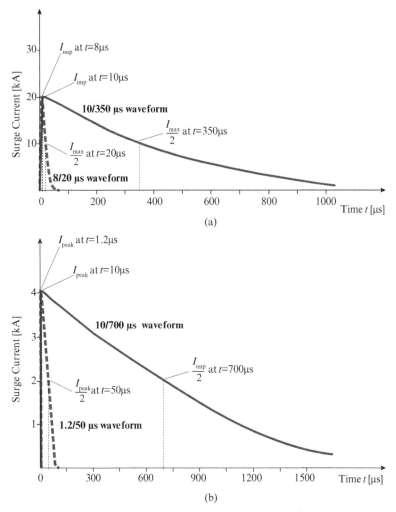

Figure 4.6 Current (*a*) and voltage (*b*) waveforms.

8/20 μs waveform to represent direct and induced lightning (short circuit) currents, respectively (see Figure 4.6a). When dealing with induced (open circuit) voltages, generally used to test SPDs, the standardized waveform 1.2/50 μs waveform is used; additionally, ITU-T K series suggest 10/700 μs voltage surges for data systems (see Figure 4.6b).

The most popular mathematical function to represent lightning waveforms has been the double-exponential expression; nevertheless, experimental data showed that return strokes have a concave rising portion with no discontinuity at $t = 0$. Nowadays, IEC 62305-1 stipulates the Heidler function as the general functional expression for the return stroke and continuing current waveforms (4.1):

$$i(t) = \sum_{k=1}^{m} \frac{I_{0k}}{\eta_k} e^{-\frac{t}{\tau_{2k}}} \frac{\left(\frac{t}{\tau_{1k}}\right)^{n_k}}{1 + \left(\frac{t}{\tau_{1k}}\right)^{n_k}} \tag{4.1}$$

with

$$\eta_k = \exp\left[-\frac{\tau_{1k}}{\tau_{2k}}\left(n_k \frac{\tau_{2k}}{\tau_{1k}}\right)^{\frac{1}{n_k}}\right]. \tag{4.2}$$

The constant I_{0k} controls the amplitude, n_k controls the initial waveform steepness, τ_{1k} is the front-time constant, τ_{2k} is the decay-time constant, and η_k is the amplitude correction factor.

The striking distance is the core concept of the *electrogeometric model* (EGM) and its offshoot *rolling sphere method*, which are commonly used to describe the protected zone of a structure where lightning cannot strike, and design the external *lightning protection system* (LPS).

The external LPS intercepts the lightning flash thanks to an air-termination system, usually comprising vertical rods/mast, horizontal wire or mesh system, conducts the lightning current safely to earth, using concealed or exposed down-conductors, and dissipates it into the ground through an earth-termination system, that is, a buried ground system composed of conducting elements or foundation earth electrodes.

Any lightning strike that penetrates the air-termination system is termed as a *direct strike* or *shielding failure*. In the EGM model, the striking distance r_s is assumed to be a function of only I_{peak} and independent of the geometry of the structure, according to the relationship $r_s = 10 I_{peak}^{0.65}$, where the peak current is in kA. This formula is used to define to radius R of the rolling sphere (see Table 4.1) in the *rolling sphere method*, introduced by the IEC 62305 to define the *lightning protection class* (LPC) offered by the LPS, which is identified by the expected risk of lightning strike and damage (lightning protection level, LPL). This method involves rolling an imaginary sphere of radius R over the surface of the structure being protected: the points that can be touched by the sphere are possible attachment points. Therefore, the external LPS (e.g., lightning masts, shield wires, fences, and other grounded metal objects intended for lightning shielding) should be properly sized and positioned to ensure that no point of the structure may come in contact with the rolling sphere. In other words, with reference to Figure 4.7a, the PV module is protected against a direct stroke because is not touched by the curved surface of the rolling sphere, thanks to the air termination rods; objects in contact with the sphere or, even worse, that penetrates the sphere's surface, are not protected.

Table 4.1 Main lightning stroke parameters for different IEC protection levels.

	I_{peak} [kA]		IEC Probability, %		
IEC Protection Levels	**Minimum value**	**Maximum value**	**For the minimum value**	**For the maximum value**	r_s [m]
I	3	200	99	99	20
II	5	150	97	98	30
III	10	100	91	95	45
IV	16	100	84	95	60

It should be observed that the effectiveness of the rolling-sphere method has been questioned for complex structures [21], and is reliable for structures of height not exceeding 60 m.

Std. IEC 62305-1 defines four different LPL, I, II, III, and IV, from the most to the less performing.

The LPL roman numerals identify the probability that minimum and maximum lightning current peaks will not be exceeded in naturally occurring lightning strikes (Table 4.1). When protection measures are adopted in accordance with IEC 62305-1, the LPS will be able to reduce damages and consequential losses, provided that the lightning strike parameters fall within the statistically defined limits of Table 4.1.

As earlier mentioned, the minimum value of I_{peak} is used to determine the radius of the rolling sphere, whereas its maximum value represents the peak current that the LPS components will have to withstand.

A second widely used method to design an LPS is the *protective angle method* (see Figure 4.7), which is a mathematical simplification of the rolling sphere method.

With reference to Figure 4.7b, the protective angle α is determined when the slope intersects the rolling sphere in such a way that the protected area (in green) and the unprotected area (in magenta) are equal.

The protective angle α can be calculated as follows [22–24]:

$$\alpha = \begin{cases} \dfrac{1}{2}\left[\arctan\left(\dfrac{\sqrt{2Rh-h^2}}{h}\right)+\arcsin\left(\dfrac{R-h}{R}\right)\right] \\[3mm] \dfrac{1}{2}\left[\arctan\left(\dfrac{R-h}{\sqrt{2Rh-h^2}}\right)+\arctan\left(\dfrac{\sqrt{2Rh-h^2}}{h}\right)\right] \\[3mm] \dfrac{180}{\pi}\left[\arctan\left(\dfrac{1}{h}+\dfrac{R}{h^2}\right)\sqrt{2Rh-h^2}-\left(\dfrac{R}{h}\right)^2\arccos\left(\dfrac{R-h}{R}\right)\right] \end{cases} \quad . \quad (4.3)$$

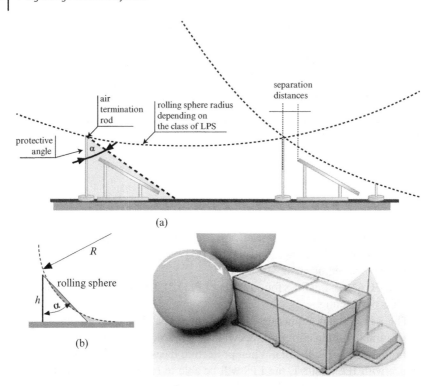

Figure 4.7 Rolling sphere method (*a*) and protective angle method (*a-b*).

In the previous equations, h ranges from 2 m to R; for $0 < h < 2$ m, the value of α remains constant and equal to the value at $h = 2$ m.

A third method is the *mesh method* that is suitable to protect buildings, regardless of their height or the shape of the roof. This method is often employed to design air-termination systems that reproduce a Faraday cage around a non-conductive wind tower nacelle[1] complemented by lightning rods and down conductors. According to this method (see Figure 4.7c), a conducting mesh with a cell size determined by the class of the LPS (5 m, 10 m, 15 m, and 20 m for LPL I, II, III, and IV, respectively), must be placed above the flat surface to be protected. The air termination conductors of the mesh must be positioned at the roof edges, on roof overhangs, and on the ridges of roof with a pitch in excess of 5.7°; no objects may protrude above the air termination system. To simplify, the sag of the rolling sphere is assumed to be zero for a

---

1 The nacelle is the housing which contains the drive-train and other elements on top of a horizontal axis WT tower.

meshed air-termination system. The application of the rolling-sphere method leads to the identification of a critical separation distance between the mesh and the structure [25].

To reduce the risk of dangerous sparking, electrical isolation between the air termination or the down conductor and the metallic parts of the installation (e.g., PV panels or inverters) must be ensured. The electrical insulation is enough when the *separation distance* between such parts is larger than the minimum value s, defined as $s = \dfrac{k_i}{k_m} k_c l$, per Std. IEC 62305-3, where k_i is a constant that depends on the selected class of the LPS, k_m is a constant that depends on the electrical insulation material, k_c is a constant that depends on the (partial) lightning current that flows in the air termination and down-conductors, and l is the vertical distance from the point at which the separation distance is to be determined up to the closest point of the equipotential bonding. For example, in a roof-top PV installation protected by rods, the distance l spans from the top level of the PV supporting infrastructure to the nearest equipotential bonding point or the earth termination system (i.e., ground level).

The safety separation distance is an internal LPS, whose fundamental role is to prevent dangerous sparking from occurring within the structure to be protected.

Another internal LPS is the equipotential bonding, such that in the event of circulation of lightning currents, no metallic part is at a different potential with respect to another conductive object. The electrical interconnection can be achieved by natural bonding or by using specific bonding conductors. Bonding can also be accomplished by the use of SPDs where the direct connection with bonding conductors is not suitable (e.g., in the case of normally live parts). Other internal *lightning protection measures* (LPMs) may consist of appropriate cable routings, cable shielding, and the installation of coordinated SPDs, typical of PV installations and wind turbines, but also of electrical systems at large. For SPDs, like the case of LPS, Std. IEC 62305-1 defines four LPLs according to the maximum expected lightning current that is statistically assumed not to be exceeded by natural lightning, which the SPD must be able to withstand.

The lightning current can be injected into the PV systems power supply mainly through three different coupling mechanisms [26] (Figure 4.8): *galvanic coupling*, *magnetic field coupling* (or inductive coupling), and *electric field coupling* (or capacitive coupling).

The prerequisite for the galvanic coupling to occur is for the lightning to directly strike a grounded exposed-conductive-parts (e.g., the metallic frames or supports of a PV module); the partial lightning current injected into the system flows to earth through the power supply lines. The impedance of the

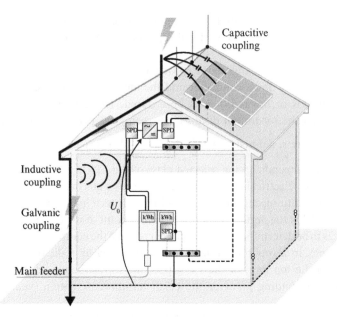

Figure 4.8 Main coupling mechanisms: galvanic, inductive and capacitive.

grounding system, when dissipating the lightning current into the ground, causes a transient ground potential rise and a voltage in the PE conductor up to several thousand volts. This coupling mechanism may also be referred to as *resistive coupling*, which is rather misleading because the grounding system must be modeled as an impendence, since the lightning is a fast transient phenomenon.

In the inductive coupling, the lightning discharge creates a time-variable magnetic field that induces voltage surges in all conductive loops of the electrical installation, especially those that involve high-impedance loads. The time-variable magnetic field is often called *electromagnetic pulse* (LEMP) due to the rapidly changing lightning magnetic field.

The *common mode* over-voltages appear between live conductors and the ground, which may cause the dielectric breakdown of the PV modules whose frame is connected to ground. The *differential mode* over-voltages appear between live conductors, which may cause the failure of electronic equipment, such as the PV inverter.

The capacitive coupling may occur between the lightning strike and the PV module frames. The electric field of a thunderstorm cloud originates a charge separation, and the electric field strength that results from the leader approaching reaches up to 500 kV/m at a distance of a couple of hundred meters to the prospective striking point, with a time variation up to 500 kV/m/μs. When the lightning discharge occurs, thanks to the established capacitive coupling, the current fiows through all conductors connected to ground as a transient surge.

4.2 Lightning Protection of PV Systems

PV systems may constitute the most prominent elements on the roof of buildings or occupy large soil surfaces. The PV installation can be *building integrated* (BIPV), when it fulfills a dual role of energy supply and roof waterproofing, shading; *partially integrated*, when it does not alter the water resistance of the roof, and *ground-based*, when it covers large areas (i.e., a PV farm). Roof-mounted PV generators do not necessarily increase the risk of a lightning strike, unless the PV system considerably increases the height of the building.

In all cases, PV systems may be vulnerable to the effects of the lightning current, which can affect the installation depending on the point of strike, especially in buildings due to wiring of the PV system inside the structure [27, 28]. PV systems may experience intense conducted and induced over-voltages in the case of lightning strikes near the installation. It has been reported that on average the 26% of damages to PV systems are caused by lightning strikes [28]. The over-voltage protection on both d.c. and a.c. sides of power electronic equipment (e.g., inverters, charge controllers, etc.) may be therefore necessary.

IEC 62305-2[2] introduces a risk assessment procedure that guides the decision-making process regarding the installation of an LPS. The standard introduces four different sources S_i of damage (Table 4.2 and Figure 4.9) and four types of damage D_i as the consequence of lightning flashes (Table 4.3). Depending on the type, use, and construction materials of the structure, each type of damage, alone or in combination with others, may produce four different type of loss L_i (Table 4.4) [29].

The damage D_i represent the "cause," whereas the loss L_i the "effect" of the lightning strike: one type of loss can be caused by more than one type of damage. It is, therefore, necessary to first identify the relevant types of loss that may occur in a structure, and then proceed to determine the related risk R_i, defined as the value of the probable average annual loss. The risk of loss of service to the public (Risk $R2$) and the risk of loss of cultural heritage ($R3$) are not generally pertinent to renewable energy systems; however, this may not always be true in the case of BIPV. The risk $R1$ of loss of a human life, and risk $R4$ of loss of economic value should instead be considered.

---

2 IEC 62305-2: *"Protection against lightning – Part 2: Risk management"*.

Table 4.2 Sources of damage.

Source	Point of strike
S_1	Flashes to the PV structure
S_2	Flashes near the PV structure
S_3	Flashes to a service
S_4	Flashes near a service

Table 4.3 Types of damage.

	Type of damage
D_1	Injury to living beings by electric shock as a result of touch and step voltage
D_2	Physical damage (e.g., fire, explosion) as a result of the physical effects of the lightning discharge
D_3	Failure of electrical and electronic systems as a result of surges

Table 4.4 Types of loss.

	Types of loss
L_1	Loss of human life
L_2	Loss of service to the public
L_3	Loss of cultural heritage
L_4	Loss of economic value

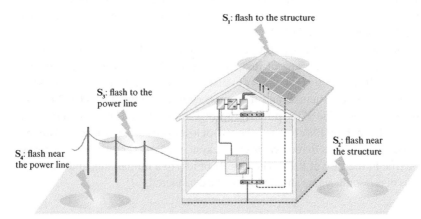

S_1: flash to the structure

S_3: flash to the power line

S_4: flash near the power line

S_2: flash near the structure

Figure 4.9 Main sources of damages.

Each risk is the sum of different risk components Rx (Table 4.5), as per IEC 62305-2, and each risk component can be evaluated as

$$R_x = N_x P_x L_x \qquad (4.4)$$

where N_x is the number of dangerous events per year; P_x is the probability that a dangerous event causes certain damage to the structure; L_x is the *loss factor*, that is, the quantitative evaluation of the effects, amount of loss, extent, and consequences of a certain damage.

According to IEC 62305-2, a multi-step risk management procedure should be followed: (i) identification of the object to be protected with its main characteristics; (ii) identification of all the types of losses L_i and the relevant corresponding risk R_i; (iii) evaluation of risk R_i for each type of selected loss;

Table 4.5 Risk components.

Types of loss	
Risk components for a structure due to flashes to the structure itself	
R_A	Injury to living beings caused by touch and step voltages
R_B	Physical damage caused by sparking inside the structure triggering fire or explosion
R_C	Failure of internal systems due to lightning flash striking the system and causing LEMP
Risk component for a structure due to flashes near the structure	
R_M	Failure of internal systems caused by lightning flash striking close enough to the system that it may cause LEMP
Risk components for a structure due to flashes to a service connected to the structure	
R_U	Injury to living beings caused by touch voltage inside the structure, due to lightning current injected in a line entering the structure
R_V	Physical damage (fire or explosion triggered by sparking between external installation and metallic parts) due to lightning current transmitted through or along incoming services
R_W	Failure of internal systems caused by over-voltages induced on incoming lines and transmitted to the system through a service connected to it
Risk component for a structure due to flashes near a service connected to the structure	
R_Z	Failure of internal systems caused by over-voltages induced on incoming lines and transmitted to the system (flashes near a service connected to the structure)

(iv) evaluation of the need for protection, by comparison of the calculated risk R_i with the tolerable risk R_{Ti}.

The loss of economic value L_4 is an exception to the general rule that calls for the comparison between the calculated risk and the tolerable risk. For this type of loss, in fact, the economic benefits of the installation of the LPS are the parameters to decide if the LPS should be installed. A cost-benefits analysis should be performed to evaluate the effectiveness of the financial investment necessary to install the LPS.

It is initially necessary to calculate: the annual cost C_L of the total economic loss in the absence of the LPS; annuitized the cost of the protection measures C_{PM} and the annual cost C_{RL} of the residual economic loss with the LPS (if the LPS is installed, the magnitude of the economic loss is reduced, but is never zero, since a residual risk is always present). If the annual saving in money $S_M = C_L - \left(C_{PM} + C_{RL} \right)$ is negative, the installation of the LPS may not be cost-effective; otherwise, the LPS makes economic sense and allows cost saving during the life of the PV installation.

If data for the above analysis are not available, IEC 62305-2 allows a value for the tolerable risk R_T of 10^{-3}/year.

In general, if the calculated risk R_i exceeds the tolerable risk R_{Ti}, the PV system should employ an LPS or/and LPMs to guarantee the person's safety, characterized by a proper LPL to reduce the calculated risk R_i below the tolerable risk R_{Ti}.

4.2.1 Ground-Mounted PV Systems

In ground-mounted PV systems, the risk of loss of human life R_1 due to flashes to or near the PV system, is generally deemed irrelevant. In such PV installations, in fact, the probability of the continuous presence of persons within 3 m of the PV modules is considered to be extremely low; in addition, the soil resistivity around the PV array may be 5 kΩ·m or greater, for example, due to a layer of gravel of thickness of at least 15 cm. In both the above cases, the probability of damages for step and touch voltages is deemed negligible. PV structures are basically incombustible; therefore, the risk for a spark to trigger fires or explosions is also deemed negligible.

Based on the above, the only risk generally relevant for renewable energy systems is the risk R_4 of the loss of the economic value (Eq. 4.5).

$$R_4 = R_B + R_C + R_M + R_V + R_W + R_Z. \qquad (4.5)$$

The loss of economic value L_4 may include damage to the PV array, inverter breakdown, loss of production of electric energy, etc. The risk components are defined in Table 4.5, and summarized in Figure 4.10.

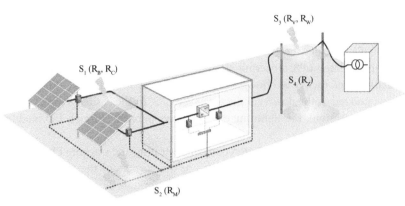

Figure 4.10 Risk components for ground-mounted PV generators.

For PV systems, the most relevant risk components to R_4 are: R_M, R_W and R_Z. The reduction of the risk R_4, if deemed cost-effective, may be achieved by installing SPDs at the service entrance to decrease R_W and R_Z, and at the d.c. side of the PV installation to decrease R_M.

SPDs, as it is later on discussed, are designed to limit transient over-voltages of atmospheric origin by temporarily diverting lightning currents to ground, so as to limit the magnitude of the overvoltage to a value that is not hazardous for the PV system.

In addition to the SPDs, external LPS (e.g., air-termination rods) may also be installed. Two types of external LPS are commonly applied, namely isolated and non-isolated LPS, as shown in Figure 4.11 [30, 31].

Isolated LPS is used when there is a high risk of damage caused by the direct lightning current. Isolated LPS consists of a free-standing mast, including the air terminal and the down conductor. The air-termination system is set up

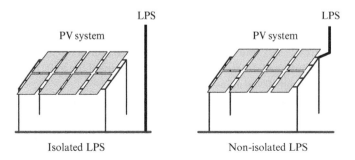

Figure 4.11 Isolated and non-isolated LPS.

observing the minimum separation distance s to prevent dangerous sparking against parts of the PV system. It is strongly recommended to maintain the separation distance between the PV modules and any metal parts.

In non-isolated LPS, air terminals are directly installed on the PV mounting structure (e.g., racks), as shown in Figure 4.11. The metallic mounting structure is utilized as the down conductor of the LPS. In the case of a direct strike to the air terminal, the lightning current flows to earth through the module supporting structure.

This practice is often encountered in concentrating and tracking PV systems. Small air-termination rods are installed at the corners of the tracking base, which capture and force the lightning current through the metal structure of the tracking base. The length of the rods should be determined according to the rolling sphere method such that the rolling sphere does not touch the tracking base for any of its possible positions.

When designing an LPS, the distance between the PV panels and the external LPS is essential to prevent excessive shading. Diffuse shadows cast by overhead lines do not significantly affect the PV arrays or their yield. Objects that cast a dark shadow negatively affect the PV modules, impacting the flow of the PV current. The appearance of an umbra and penumbra on the solar modules should always be avoided: *umbra* is the dark area where the irradiance is reduced to zero, while the surrounding shaded area is the *penumbra*, where the irradiance may be just reduced. Generally, LPS should be placed at the North side of PV arrays, and the lightning rods be installed at a distance l greater than a minimum distance l_f from the PV modules; l_f can be calculated with the simple relationship $l_f = 108 d_f$, where d_f is the diameter of the lightning rod.

4.2.2 Rooftop Mounted PV Systems

The average annual number N of dangerous events resulting from lightning strikes affecting a specific building depends on the thunderstorm activity of the geographic area where the structure is located, on the characteristics of the area itself (i.e., the presence of surrounding buildings, the type of terrain, presence of trees, etc.), and on the edifice's physical characteristics, such as its height. The function of the risk analysis is to determine if the building needs an LPS by considering all the above factors.

If a building has no LPS, the installation of rooftop mounted PV modules may increase N by changing the overall building's height (e.g., rack-mounted PV generators on flat roof). As a result, the risk of damage for the entire building may exceed the tolerable value. In this case, a lightning risk analysis for the entire building must be performed, to establish the risk of loss of human life R_1 and, also, the risk of economic loss R_4. With reference to Figure 4.12, the risk components for the loss of human life L_1 are expressed in Eq. 4.6.

$$R_1 = R_A + R_B + R_U + R_V \qquad (4.6)$$

If a lightning protection system is already installed on top of the building, the rooftop-mounted PV array and associated equipment should be within the protection zone of the LPS; if the PV array is not within the protection zone of the LPS, additional air terminations (e.g., lightning masts) may need be installed, usually maintaining the separation distance s. When the separation distance cannot be provided (e.g., PV array on a metal roof), the components of the PV system must be connected, directly, or via SPD, to the existing air-terminations and down-conductors, which connect the air-terminations to the grounding system.

When the separation distance s cannot be guaranteed, another solution, which may be found when hazardous areas inside or near the building are present, is the use of high voltage insulated (HVI) down conductors. These conductors feature a special insulating material and sheath for the cable inner core, which provide an equivalent separation distance from other conductive parts of the building, electric lines, and pipes. HVI conductors offer an effective and aesthetically pleasing solution on projects where the separation distances are difficult to achieve. HVI conductors are also suitable for hazardous areas.

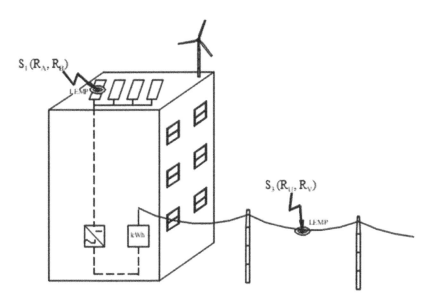

Figure 4.12 Risk components for rooftop-mounted PV generators.

Active lightning protection systems are of particular interest for renewable energy systems [32].

Early Streamer Emission (ESE) lightning protection systems create ionized electrical fields that increase the probability of lightning anticipation. An ESE is equipped with an active device that generates pulses of controlled magnitude and frequency at the tip of the terminal during a storm, when propagation field conditions are favorable, prior to a lightning discharge. This increases the probability of initiating an upward streamer that connects with the downward propagating leader. This enhancement should apply to both polarities of natural lightning, although discussion of the polarity dependence of ESE systems is currently not present in literature. Commercial ESE systems generate only positive ions, which attract the negative heavy electrical loads in the clouds; thus, they are effective only under negative cloud-to-ground lightning conditions. ESE systems can create a safety dome of a particular radius called a "protection area," whose size is based on a parameter termed the "advance time" ΔT. According to Std. NFC 17-102/2011, ΔT must range between 10 and 60 μs, which allows the ESE system to protect an area ranging between 40 m and 150 m.

Charge Transfer Systems (CTS) (also called lightning suppressors) [33] prevent a lightning strike from occurring within a protected area, rather than favor it. While the ESE system produces ionized air to create an artificial electrical field, the CTS collects the charge developed by the clouds from the protected area and transfers it into the surrounding air.

CTS sends out electromagnetic pulses, which decrease the electric field between the cloud and the ground, which is the trigger for a lightning strike, which can no longer form (with a claimed success rate of 99.9%).

It should be observed that when using these systems, lightning strikes could be only diverted to neighboring structures.

4.2.3 Protection against Overvoltage

Direct or nearby lightning strikes are prone to hit PV power systems during thunderstorms. Strikes falling in the vicinity of PV systems, more frequently than direct strikes [34], have caused large transient overvoltages (TOVs) with non-negligible damages to the PV circuitry and structures [35], and module degradations [36]. TOVs can be classified in slow front overvoltages (20 μs $\leq T_1 \leq$ 5 μs, $T_2 \leq$ 20 ms), fast front overvoltages (0.1 μs $\leq T_1 \leq$ 20 μs, $T_2 \leq$ 300 μs), and very fast front overvoltages ($T_1 \leq$ 0.1 μs).

Generally, computational electromagnetics is applied to model the interaction among electromagnetic (EM) fields and objects by solving Maxwell's equations with appropriate boundary conditions [26, 37, 38]. Numerical methods for analyzing LEMP transients in conducting structures can be broadly classified according to two methodologies: transmission-line models and full-wave models. The former method couples the EM field radiated by the return stroke, modelled as a travelling wave along a straight antenna (referred to as the *engineering model*), with a multiconductor transmission line whose equations assume the propagation of a transverse electromagnetic (TEM) wave [36]. The coupling is done through distributed voltage and/or current generators according to well-established theories [39]. In contrast, the full-wave methods model the EM coupling mechanisms in soil and air at the price of a much more complicated numerical procedure. Full-wave models include the numerical methods developed today, e.g., method of moments (MoM) [40], finite element method (FEM) [41], partial element equivalent circuit (PEEC) [27], and boundary element method (BEM), both in frequency and time domains, and the finite-difference time-domain (FDTD) [42] method as well.

The goal of the over-voltage protection is ultimately to prevent the failure of the basic insulation of live parts (under a differential mode TOV) or between live parts and ground (under a common-mode TOV) of the PV system, and therefore permanent damages to the equipment. To this purpose, the rated *impulse withstand voltage* U_w, assigned by the manufacturer to the PV equipment, may be used to verify if its insulation withstands capability is greater than the common mode overvoltage. The impulse withstands voltage U_w is the capability of equipment to withstand overvoltage surges and is classified according to 4 categories, from I (sensitive equipment) to IV (industrial equipment), as per IEC 60664-1.

According to IEC 62305-2, flashes striking near buildings do not endanger the human life (Risk R_1), except in structures with a risk of explosion and hospitals, where failures of electronic systems may immediately threaten patients. Therefore, the only risk generally to be considered for PV systems is the loss of economic value (Risk R_4, where the predominant risk component is R_M), due to damages to the PV array, the inverter, and due to the loss of production of electric energy. It should be noted that a European study[3] indicated that the chance for an inverter to sustain damages from nearby lightning strikes is once in 40 years.

To decrease the magnitude of induced overvoltages on both a.c and d.c. circuits, the induction loop area must be minimized by an appropriate line routing and/or using shielded cables, bonded at least at either end. On both the a.c and d.c. circuits of PV systems, cables should be twisted together to reduce any loop surface. These *routing precautions*, whose goal is to minimize the

---

3 Task 7 Report IEA-PVPS T7-08: 2002: *"Reliability Study of Grid Connected PV Systems"*.

probability that a flash near the PV system causes the failure of its components, is effective if the induction loop area formed by either the d.c. or a.c. cables does not exceed 0.5 m^2, as per IEC 62305-2. The minimum value for the loop area is, however, dictated by the mechanical constraints imposed by the pre-installed junction box on the backside of the solar module (Figure 4.13). The cable spacing at the panel junction box is, in fact, around 8 cm, and the cables have a length of 90 cm, which limit the reduction of the induction loop area obtainable with routing precautions.

The adoption of routing precautions on the a.c. side may consist of keeping the protective conductor as close as possible to the line conductor, installing it in the same conduit, or specifying it as a part of the same cable.

To avoid dangerous sparking, all the metallic parts that are not separated more than the separation distance s, must be connected through an appropriate equipotential bonding, such that in the event of lightning currents flowing, no metal part is at a different potential with respect to another. The *bonding measure* requires that the metallic parts, the incoming services, and extraneous conductive parts shall be bounded to a single earth reference point (i.e., the equipotential bonding busbar) that is connected to the earthing system to form a complete integrated meshed bonding network, which is often encountered in ground-mounted PV systems.

4.2.4 Surge Protective Devices (SPDs)

If lightning-induced overvoltages cannot be sufficiently reduced by shielding, routing, or bonding precautions, the rated impulse withstand voltage U_w of PV equipment may not be sufficient to withstand surges possibly occurring between live conductors and the ground. In this case, to protect the PV equipment, Surge Protective Devices (SPDs) may be installed according to Std. IEC 60364-7-712.

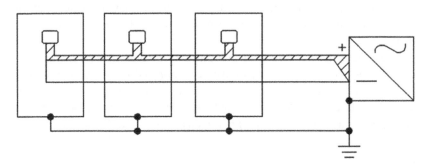

Figure 4.13 Minimum induction loop area.

There are three types of SPDs:

- Type 1 SPDs are designed for protection of electrical installations against transient overvoltages (TOVs) due to direct lightning strokes. The main test parameter is the *lightning surge current* I_{imp}, which is the peak value of the current flowing through the SPD with pulse shape 10/350 µs that the SPD can discharge five times without destroying itself. The peak value I_{imp} that is used to represent partial lightning currents is related to the maximum values of discharge currents that are expected to occur at the LPL probability level at the location of the installation of the SPD in the system (e.g., 25 kA per pole – 100 kA per 3P + N system, for LPL I). Spark gap devices [43], not usually encountered in PV installations since they cannot extinguish dc arc currents, are also characterized by the *auto-extinguish follow current* I_{fi}, which is the current that the SPD can cut after the overvoltage activation, and that must be higher than the short circuit capacity at the installation point.
- Type 2 SPDs are designed for protection of low voltage electrical installations against TOVs due to switching and indirect lightning strokes. They are characterized by two parameters: the *maximum discharge current* I_{max}, which is the peak value of the current flowing through the SPD with pulse shape 8/20 µs that the surge protector can withstand without destroying itself, and the *nominal discharge current* I_n, which is the level of impulse current a surge protector can divert to earth repeatedly at least 20 consecutive times, without destroying itself. The difference between I_{max} and I_n indicates when the SPD is working near its limits in nominal operating conditions. In other words, the higher the value of I_{max} is for the same I_n, the safer the SPD is working, away from its performance limits.
- Type 3 SPD is designed for local protection of sensitive loads (e.g., electronic devices, hard-wired devices, telecoms and data devices, string surveillance transducers). These low discharge capacity SPDs provide fine protection against 8/20 µs surge currents and against 1.2/50 µs TOVs lightning striking occurring far away. They are mainly characterized by the *open circuit voltage* U_{OC}, which is the crest value of the off-load discharge voltage surge delivered by a test 1.2/50 µs waveform simultaneously with an 8/20 µs short-circuit current waveform (combined voltage/current impulse test) that the SPD can safely handle. The higher this value, the greater is the capacity of the protection circuit to handle surges. Generally, a level of 2.5 kV or more (4 kV and 6 kV) is recommended. Type III SPDs should always be installed in conjunction with type II SPDs.

Any SPD has a *residual voltage* U_{res}, which is the value of the voltage at its terminals when the SPD is subject to the flow of an impulse current; the residual voltage is a function of the amplitude and waveform of the discharge current.

When the discharge current is the impulse current I_{imp} (for Type 1) or the nominal current I_n (for Type 2), the residual voltage is termed the *level of protection* U_p, which is a key parameter for the design and choice of the SPD.

Finally, the protection voltage U_{prot} is the sum of the level of protection U_p and the inductive voltage drops U_L across the SPD connections, i.e., $U_{prot} = U_p + U_L$. The protection voltage U_{prot} must always be less than the impulse withstand voltage U_w of the equipment to be protected. As a general rule, the per unit length (p.u.l.) lead inductance L' of SPD's connections is assumed to be 1 μH/m; the inductive voltage drop is approximately 1 kV/m of lead length when it is caused by an impulse current with a rate of rise of 1 kA/μs.

When dealing with SPD, it has to be borne in mind that inductive effects always play a major role, as the lightning discharge is a fast transient phenomenon. If the distance between an SPD and the equipment to be protected, referred to as the *oscillation protection distance* l_{po}, is too large, oscillations could lead to a voltage at the equipment terminals, which may reach a magnitude of twice the protection level U_p. This can cause a failure of the equipment to be protected, in spite of the presence of the SPD; therefore, a satisfactory *protective distance* must be assured. Generally, the protective distance depends on the SPD technology, the type of system, the rate of rise of the incoming surge, and the impedance of the connected loads [44, 45]; such distance can be studied by means of the travelling waves theory on multiconductor distributed lines. Practically, when the protective distance is kept under 10 m or $U_{prot} < U_w / 2$, voltage oscillations may be ignored; when either condition is not satisfied, the oscillation protection distance can be estimated as $l_{po} = (U_w - U_{prot}) / k$, where the factor k is equal to 25 V/m [35].

The inductive effect is made worst by the fact that lightning flashes can induce an overvoltage in the circuit loop comprising the SPD and the equipment, decreasing the protection efficiency of the SPD. In this regard, the induction protection distance l_{pi} is the maximum distance between the SPD and the equipment for which the protection of the SPD is adequate. Avoiding large loop surfaces or line shielding can reduce the effect of the voltage induction, which would allow to ignore the induction protection distance. Otherwise, l_{pi} can be estimated as $l_{pi} = (U_w - U_{prot}) / (3000 K_{S1} K_{S2} K_{S3})$, where K_{S1} and K_{S1} are the factors of the shielding effectiveness at the boundary between *lightning protection zones* (as explained later on) 0–1 and 1–2, respectively, and K_{S3} is the factor of routing precaution on the wiring.

In any PV installation, both in the d.c. and a.c. side, SPDs should be installed close to the equipment to be protected (PV array, power conversion devices, string surveillance transducers, etc.). When a very high protection level is needed, more SPDs can be connected together, creating a combined surge protective device that is an all-in-one solution (e.g., type 1 + 2 commercial SPDs).

SPDs for PV circuits must be designed specifically for d.c. applications, since a.c. SPDs upon failure can disconnect their circuit but cannot quench the d.c. arcs, and can cause fires. A.c. SPDs, in fact, rely upon the voltage zero crossings to extinguish the arcs, which are obviously not present in d.c. circuits.

Today, most of the SPDs are based on the gapless surge diverter technology, initially introduced in metal oxide varistors (MOVs). They exhibit a voltage-dependent impedance: in normal operating conditions their impedance is very high (i.e., an open circuit), but in the presence of a transient overvoltage, their impedance decreases, and the associated current is safely diverted to ground, while the residual voltage across their terminals is kept within a specified value. SPDs are developed by mixing zinc oxide (ZnO) with additive materials and the electrical properties are controlled by their grain boundaries: they are characterized by fast transient voltage switching and have better long-term performance due to their high-energy absorption capabilities.

Standard IEC 61643-32 makes a distinction between voltage-limiting SPDs based on varistors, and voltage-switching SPDs based on spark gaps, or gas-filled surge protective devices. Spark gap devices, as previously mentioned, are not the best choice for d.c. circuits, as once their conduction has been triggered by the overvoltage, they may not be able to interrupt it until the voltage across their terminals is typically less than 30 V. The d.c. operating voltage of PV modules, typically greater than 30 V (and with no zero crossings), may be higher than the arc voltage that occurs across the spark gap terminals during the conduction; thus, the arc voltage may persist even if the surge has expired, keeping the live wires grounded. Thus, the preferred type of SPD for the protection of PV arrays is the MOV type.

MOV SPDs, when installed in the a.c. side PV systems, are usually equipped with fuses, which can interrupt the current flow in case of a fault, and to put the device in a safe condition.

In PV applications, SPDs protect PV arrays against indirect lightning and power conversion equipment, both in the d.c. and the a.c. side. SPD should be installed between live wires and ground.

Commercially, SPDs with Y configuration are commonly used, especially in utility-scale PV systems. In this configuration, each leg of the Y contains an MOV module. In an ungrounded system, the two upper modules of the Y are connected between the live positive and negative polarities and a central node (*differential mode* protection), and the lower module between the central node and the ground (*common mode* protection). Each module is rated for half the system voltage, so that even if a pole-to-ground fault occurs, the MOV modules do not exceed their rated values.

If the negative pole is grounded, the same SPD equipment can be maintained without connecting its negative pole (the negative wire is grounded and is not considered a live conductor).

When used to protect PV arrays from over-voltages caused by indirect lightning, the type 2 SPDs are usually located in the combiner boxes (Figure 4.14), and their recommended specifications are:

a) maximum continuous operating voltage $U_C > 1.3 \times V_{OC\,STC\,GEN}$
b) maximum discharge current $I_{max} \geq 5\,kA$. This is a conservative requirement because, according to Std. IEC 62305-1,[4] expected surge overcurrents due to lightning flashes near a structure do not exceed 200 A.
c) voltage protection level $U_C < U_p$ (at least 1.1 kV).

Due to the characteristics of PV panels, fuses installed in series to d.c. SPDs are usually ineffective. The short circuit current from a typical PV array has a limited magnitude, very close to the maximum power point current, which may not allow the operation of the fuse. For this reason, SPDs must be chosen with a short circuit withstand current I_{SCW} greater than the short circuit current of the PV array $I_{SC,PV}$ to which the SPD is connected.

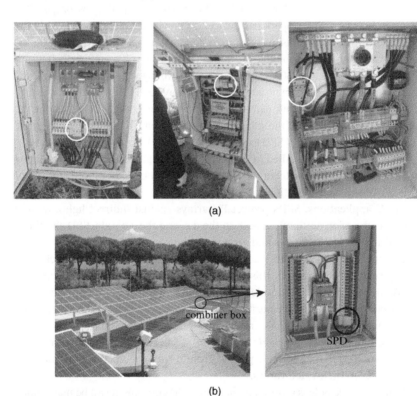

(a)

(b)

Figure 4.14 SPD in combiner boxes.

---

4 IEC 62305-1: "*Protection against lightning –Part 1: General principles*" 2012.

As to the protection of inverters against overvoltages, an SPD type II is usually sufficient. If partial lightning currents are expected, an SPD type I with connected SPD type II should be used. Commercial PV inverters may be already fitted with internal SPDs at the d.c. input terminals; however, the installation of external SPDs may still improve the protection of the PV system, and possibly prevent the internal SPD from causing the inverter to go in stand-by, which would then require qualified personnel to restore its operation. The two SPDs, however, should be properly coordinated; the characteristics of the internal SPD of the inverter may not be known, which would prevent an effective coordination.

SPDs within inverters may cause damages due to interaction with the EMC filter, and, in the case of excessive voltage, the high current within the overvoltage protection device may also impact the inverter electronic circuits.

The d.c. SPD should be installed in the vicinity of the inverter. If this is not possible, induced overvoltages may occur along the loop between the SPD and the inverter, whose magnitude increases with their separation distance. The number of SPDs required to protect the PV system is based on the distance between the PV array and inverter. If the distance between the PV array and the inverter is less than 10 m, a single SPD installed as close as possible to the inverter, will suffice. Otherwise, if the distance between PV array and inverter is greater than 10 m, two SPDs should be installed, one close to the inverter and the other close to the PV array. In large scale PV systems, additional SPDs should be connected next to the PV array (in the combiner boxes as previously shown) to provide protection even with the d.c. disconnect open.

When using string fuses and only one SPD, as typical within junction boxes of utility-scale PV systems, the SPD must be installed at the point of interconnection of the combined strings after the fuses. If the SPD were connected to only one string, between the string input and the string fuse, the remaining strings would be unprotected, if the fuse tripped. For inverters with multiple MPP trackers, SPDs should be installed for each input. The same applies to inverters with only one MPP tracker but multiple inputs, each with its own fuse (Figure 4.15).

An SPD should be installed also on the a.c. side of the inverter and be positioned as close as possible to the origin of the a.c. supply, usually at the main distribution panel, unless the distance between inverter and the board is greater than 10 m. In that case, two SPDs should be installed, one close to the inverter and the other close to the panel. As opposed to the d.c. side, multiple inverters can be protected by only one SPD on the a.c. side, since they are connected to the same grid. The integration of a.c. side SPDs is not usually planned by constructors, since multiple inverters are mounted next to each other. The separate installation of a single overvoltage protection device for all inverters is, in this case, significantly more cost-effective. The a.c. SPDs should be correctly designed for the a.c. system neutral configuration and should be protected by fuses or circuit breakers, as a back-up protection.

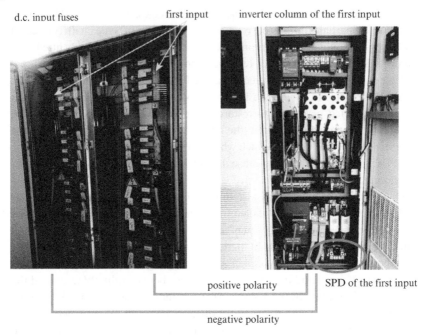

d.c. input fuses   first input   inverter column of the first input

positive polarity   SPD of the first input

negative polarity

Figure 4.15 SPD in a central inverter station.

The above general rules must be applied also based on the presence of an external LPS. If a roof-top PV system is not protected by an external LPS, two sets of SPD type 2 are needed on both the d.c. side and the a.c. side and the main panel, assuming a distance of more than 10 m. With an external LPS, if the separation distance s is maintained between air terminations and bonding conductors connected to modules' frame and mounting system, the PV system can be kept isolated; in this case, the same sets of SPDs are necessary as before, but the SPD in the switchboard should be of type 1. For non-isolated PV systems, where the LPS is connected to the modules' frames to keep the distance s, the two SPDs on the d.c. side shall still be of type 2, but the two SPDs on the a.c. side shall be of type 1. A designer guidance for planning and implementing SPDs for low voltage power systems can be found in [9–11].

4.3 Lighting Protection of Wind Turbines

The impressive growth of the wind turbines (WT) industry features tens of thousands of new tall structures, averaging heights of over one hundred meters, installed onshore on flat and/or elevated locations and offshore in

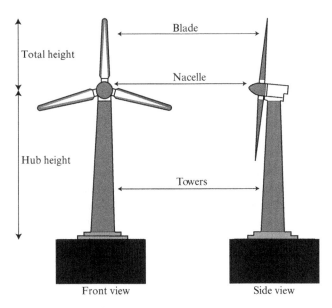

Figure 4.16 Three-blade, horizontal axis wind turbine.

open waters (Figure 4.16). Convenient sites for WTs may be regions of intense thunderstorm activity, which causes the WTs to be exposed to direct lightning strikes; this can cause harm to service personnel, economic losses due to damage, and loss of revenue. It has been stated that 30% of the damages to WTs are caused by direct lightning strikes, and the remaining 70% by indirect lightning [14]. According to IEC 61400–24, old WTs frequently experienced damage to the control system, whereas new WTs experience relatively more often damage to the blades. This switch of the damage location may be due to the improved transient protection of control systems and the increase in the size of WTs. Lightning protection of WTs presents peculiar problems [46], and not only because they are tall structures and frequently placed at prominent locations exposed to lightning. The most exposed WT components, such as the blades and the nacelle, may be made or covered of composite materials, e.g., carbon-reinforced-plastic (CRP), which can neither sustain direct lightning stroke nor conduct the intense lightning current.

Both blades and nacelle rotate according to the wind speed and direction, and the movement may considerably increase the number of strikes to which the WT may be subjected. The lightning current may flow through the WT structure to ground, passing through or near practically all WT components. The magnitude of induced electromotive forces is made worst by the fact that the lightning current injected into the turbine is affected by reflections at the top, bottom, and the junctions of the blades with the static base of the turbine.

WTs in wind farms are electrically interconnected and often placed at locations with poor soil resistivity. Additionally, modern WTs are characterized by the presence of ever-increasing amounts of control and processing electronics. Most of the strikes to modern turbines are expected to be upward lightning, as previously described. Neglecting upward flashes might result in an important underestimation of the actual number of strikes to a tall WT. The presence of carbon-reinforced plastic introduces new problems related to the way these laminates are bonded to other conducting components. Eddy currents in CRP laminates cause significant energy dissipation, which might result in mechanical stress on the blades. The design of the LPS of modern WTs is, therefore, a challenging task [14, 46–49].

In the event of lightning strikes, persons in and around turbines may be subjected to hazardous step and touch voltages, or explosions and fires caused by lightning flashes.

The erection and commissioning of a WT may take a few weeks, and during this time workers are in, on, and near the structure (Figure 4.17). Thus, lightning strikes put at risk personnel working on the nacelle and the blades, or simply standing next to the tower. During lightning activity, personnel should be instructed to stop working and go to safe locations (e.g., inside the WT's tower) until the hazard is over. Safety measures should be in place during

Figure 4.17 Pictures taken during the construction of a wind tower.

construction: cranes, generators, etc., should be connected to a grounding system at the early stage of activities.

4.3.1 Lightning Protection System (LPS)

To fulfill the safety requirements for personnel, for the protection of equipment and reduce possible downtime, WTs must be equipped with a *lightning protection system* (LPS). Lightning protection mitigates lightning damage by providing to the lightning current a low-impedance ground path away from equipment and structure (see Figure 4.18). The LPS must facilitate the attachment of the lightning strike to a preferred point, such as the air termination system on the blade. This allows the flow of the lightning current to earth without causing damage to systems, including the damage resulting from high levels of electric and magnetic field, and the minimization of voltages and voltage gradients observed in and around the WT.

The LPS must protect two sub-systems unique to WTs: the rotor blades and the drivetrain hosted in the nacelle (i.e., turbine, gear box, and electric generator). Lightning damages may include the breakdown of blades, of mechanical and electrical parts, and of the control systems. Repair or replacement of the blades, which are the most exposed to damages from the electric field associated with the lightning, result in the longest outage time.

In general, the lightning protection for WTs involves an *exterior* and an *interior* lightning protection system. The *exterior* lightning protection system consists of a low resistance path (e.g., down conductors) from the blade tips to

(a)   (b)

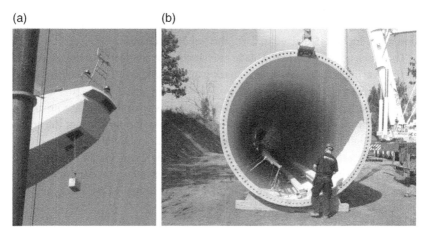

Figure 4.18 Lightning protection system over the nacelle.

the tower grounding system (e.g., concrete-encased electrode, ground rod, ring, etc.), which assures a safe dissipation of lightning currents. The yaw-bearing[5] and the tower are bolted together and, therefore, the tower may be used as a down conductor to carry the lightning current to the tower base flange, which is connected to the grounding system.

The *interior* lightning protection system includes measures against the effects of electromagnetic fields induced by the lightning current onto the electrical equipment. Lightning down-conductor(s) and adjacent power and control conductors, within the tower and nacelle, must not form large loops, to avoid large hazardous induced voltages. Through the equipotentialization of internal metal parts and external conductors entering the WT, the interior lightning protection system can prevent the build-up of electrostatic charge and therefore dangerous sparks from occurring.

The *interior* lightning protection may also include surge protection devices (SPDs) within electrical cabinets as a further protection of equipment.

As previously explained, the *attachment point* is the origin of the upward stream of charge, and the WT's blades, generally significantly taller than adjacent objects, are the most likely attachment points for lightning strike. A lightning air termination system is therefore installed on the blades of modern WTs. This system can be of different types: receptors or metal meshes on the side of blades, high resistive tapes and diverters, as well as conducting materials for the blade surface around both leading and trailing edges.

An LPS that is widely used for blades is a system of external receptors in conjunction with internal down conductors placed inside the blades. Metal receptors, which act as air terminations and penetrate the blade surface, are installed at the tip of the blades and on the side with interspace ranging between 5 and 8 m for larger blades; the exact number of receptors on each blade depends on the manufacturer's choice. All receptors are connected to internal down-conductors [50] sized to carry the lightning current.

The down conductors of the blade are usually not connected to the base plate of the bearing but, by means of a spark gap, directly to the nacelle, thus completely bypassing the pitch bearing and hub (Figure 4.19a, b). In this way, the metallic parts of the blades connected to the down conductors are normally insulated from ground, unless a sufficiently high potential triggers the spark gap. The lightning current is transferred from the metallic sheet located at blade root to the gutter ring placed on the nacelle by means of a no-contact system (referred to as the *hammer*), which consists of a non-conductive arm with two metallic receptors at its tip. The arm is bolted to the blade bearing

---

5 The yaw system allows the rotation of the rotor about the vertical axis.

outer ring, so that the receptors are faced with the blade band and with the gutter ring; in between, there is a separation ranging approximately between 5 to 40 mm. The gap between hammer receptors and metallic sheets is designed to assure that an electric arc appears between hammer tips and both conduction bands (blade and nacelle) in case of a lightning strike coming from the blade. This design avoids lightning current from flowing through the hub.

According to NREL, rotor blades with built-in conductors are far less likely to experience extensive damage, compared to those without them [51]. Both field observations [48, 52] and laboratory experiments [53] have shown that the receptors are the most exposed part of the turbine, but are not the only places where lightning may attach to the blade; the region of the blade away from the tips and toward the blade root may also be struck. Laboratory experiments have shown lightning attachment points in the middle of the blade, with arcs burning along the blade surface or penetrating into the blade boards [53]. Current LPS for blades are designed to withstand 98% of lightning strikes [54], but there is still a risk of damage, particularly at the attachment point. Lightning can damage the blades in many ways, e.g., surface discharge, breakdowns, and punctures.

The nacelle structure, like the blades, is also susceptible to initial attachment and must be part of the lightning protection system. The nacelle must be able to withstand lightning flashes according to the assigned lightning protection level. The best choice is to design it with metal covers or metal reinforcements, connected to the down conductors, to obtain a sort of metal shield that is closed in itself. With this design, a protected volume can be obtained inside the nacelle, with a considerably attenuated, electromagnetic field compared to the outside. When the nacelle is non-metallic, an air-termination system must be designed: a rear-mounted mast should be generally used, but it is highly advisable that a catenary conductor be installed for the protection of personnel working inside (i.e., a mast forward and aft with a wire hanging between). Each end of the catenary should be effectively bonded to the down-conductor system.

Lightning rods located at the rear part of the nacelle have also the crucial function of protecting all the devices and sensors installed outside the nacelle (e.g., sonic, anemometer, and wind vane). In case of a lightning strike, the current will be conducted from the air termination system to the main frame by the bolted joints of metallic platforms where the external sensors are installed, to the nacelle structure (Figure 4.19c, d). These bolted joints are reinforced by electrical joints.

It is always necessary to avoid lightning currents through the yaw bearing. For this purpose, Cu/Zn conductive blocks are usually installed at the contact surface between the main frame and the yaw system. These conductive blocks

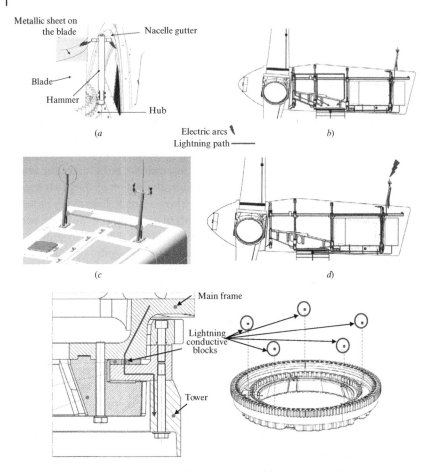

Figure 4.19 LPS of a WT: (a) Lightning path from the blade to the nacelle by the hammer; (b) lightning current path through the nacelle structure; (c–d) Nacelle rear lightning rods: Lightning current path through the nacelle struct; (e) conductive blocks location (right) and cross section of the yaw system (left).

offer to the lightning current a path of lower impedance than that of the bearing (Figure 4.19e), which is therefore protected.

According to the Std. IEC 61400-24, the metal tubular tower is the primary earth conductor and equipotential bonding connection, and therefore is the main down conductor for lightning and fault currents. However, national standards require that the upper part of the tower be directly connected to the grounding busbar at the bottom of the tower. This is the reason why copper wires may be installed along the whole length of the tower. The tower is generally erected by joining different sections by means of flanges and bolted joints. The bolted joints do allow a good electrical connection between the flanges, but to create an effective Faraday cage and guarantee equipotential bonding connection, insulated copper conductors usually connect one tower section to the other.

4.3.2 Step and Touch Voltages

The protection against the effects of lightning strikes is achieved by providing a low-impedance path to ground (e.g., down conductors), so that the lightning current can flow away from components susceptible to lightning damage. Nonconducting towers should be protected by air terminals and down conductors connected to a grounding system.

The lightning current flowing through the ground creates a potential gradient in the area of the turbine ground electrode and the remote earth. At the time of the lightning discharge, large prospective step voltages may develop in proximity of the turbine grounding system, across points of the soil surface at distance of 1 m. The step voltage decreases with the increasing distance from the ground electrode, but the hazard may be present up to a distance of 10 or 20 m.

A person touching any conducting part of the WT during a lightning strike will also be subjected to a touch voltage, particularly dangerous as the current pathway includes the cardiac region.

Let us consider a person in contact with a point on a down conductor at height h above ground (e.g., 2 m), which is carrying the lightning current I. The person establishes a loop and may be subjected to a dangerous induced touch voltage V_T (Figure 4.20). L is the inductance of the loop formed by the person touching the down conductor, standing at 1 m from it, the portion of length h of the down conductor itself and the ground. R_D is the resistance of the down conductor; $Z = Z_1 + Z_2$ is the equivalent ground impedance of the turbine ground electrode, where the connection point between Z_1 e Z_2 represents the point on the earth's surface where the person is standing; R_{Bi} is the person's internal body resistance, since at high frequency, the capacitance of the skin becomes a short circuit; Z_{BG} is the person's body impedance-to-ground.

The induced effective touch voltage V_T can be calculated as:

$$V_T = k\left(L\frac{\Delta I}{\Delta t} + R_D I + ZI\right)\frac{R_{Bi}}{R_{Bi} + Z_{BG}}, \tag{4.7}$$

where k is the fraction of the lightning current that flows through the down conductor (if more than one is present), and $\Delta I / \Delta t$ is the rate of rise of the lightning current.

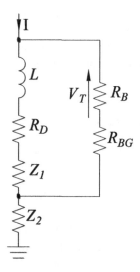

Figure 4.20 Equivalent circuit in the case of contact with the down conductor during lightning discharge.

For non-conducting towers, the reduction of the touch voltage is obtained by increasing the number of down conductors to a minimum of two. Multiple down conductors reduce both the peak and the rate of rise of the lightning current.

Z_{BG} can be beneficially increased with a layer of insulating material around the down conductors; a layer of asphalt of 5 cm thickness or a layer of gravel approximately 15 cm thick generally reduce the hazard to a tolerable level. Z_1 can be reduced with more efficient ground electrodes (e.g., meshed ground grids).

4.3.3 Lightning Exposure Assessment

The risk assessment of the average annual number of dangerous events N_D due to flashes to the WT can be performed according to the standard IEC 62305-2, introduced in the previous chapter, which is based on the determination of the *equivalent collection area* A_D [55]. A_D is the surface area at the ground level that would have the same number of annual lightning flashes as the wind turbine. The WT is supposed to be on flat ground, and the collection area is calculated in the assumption that the WT is not present. A_D is defined by the intersection between the ground surface and a straight line with 1/3 slope, which passes and touches the upmost part of the turbine and rotates around it (Figure 4.21).

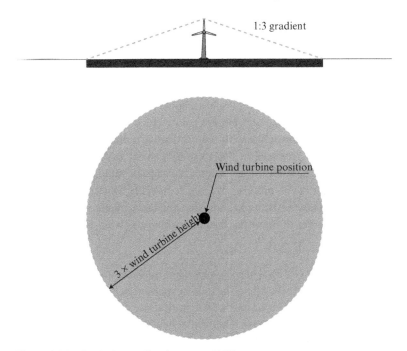

Figure 4.21 Equivalent collection area of WT.

According to IEC 61400-24,[6] the upmost part of the turbine is identified by the *effective height* H, which is equal to the hub height plus one blade length. The equivalent collection area is, therefore, a circle of radius three times H, i.e., $A_D = \pi(3H)^2$. This applies to WTs regardless of the material of the blades (e.g., blades made from non-conductive materials).

For a more theoretical definition of the collection area A_D it is possible to refer to the *attractive radii models* [56–60] that establish a relationship between the collection area of the wind tower, also referred to as *attractive area* A_D, its *effective height* H_{eff}, and the lightning current magnitude I which interacts with the WT.

To better explain this approach, we initially introduce the probability density function $f(I)$ of the lightning current amplitude I, which can be approximated by a log-normal distribution (4.8).

$$f(I) = \frac{1}{\sqrt{2}\pi I \sigma_{\ln I}} \exp\left[-\left(\frac{\ln I - \ln I_\mu}{\sqrt{2}\sigma_{\ln I}}\right)^2\right], \tag{4.8}$$

---

6 IEC 61400-24: 2019 "Wind energy generation systems - Part 24: Lightning protection".

where $\sigma_{\ln I}$ is the standard deviation of $\ln I$ and I_μ is the median value of the lightning current distribution. Plenty of literature exists on statistical values of lightning current amplitude [61, 62].

We define the *attractive radius* r_a, which describes the ability of the WT to attract lightning flashes [56]. By assuming that the lightning leader is perpendicular to the ground plane, it is generally accepted that the flash will strike the tower if its prospective ground termination point, which is the point of strike in the absence of the wind tower, lies within the attractive radius. Different expressions of r_a have been proposed in literature by several authors [56–59] as a function of current amplitude I (also referred to as peak value I_{peak}) and H_{eff}: for instance, Borghetti et al. propose $r_a\left(I, H_{\text{eff}}\right) = 0.028 H_{\text{eff}} I + 3h^{0.6}$ [56].

The relationship between the collection area and the attractive radius, which determines the number of lightning flashes striking the wind tower, is given as $A_D = \pi \int_0^\infty r_a^2(I) f(I) \, dI$ [63].

The effective height H_{eff} [64] of the tower takes into account its location, for example, on elevated terrain, where the tower experiences greater incidence of lightning flashes than that in the case of flat ground site. Wind turbines placed on a hilltop of height H_m are given an effective height H_{eff} that is often significantly larger than their actual height H, but usually lower than $H + H_m$. Several expression have been proposed for H_{eff} [65, 66].

Wind towers may be within a distance $3H$ from other turbines or tall objects; in this case, the effects of their relative location on the collection area can be taken into account by means of the *location factor* C_D (Table 4.6). Higher objects within a distance $3H$ from the WT have larger collection areas, and therefore may beneficially intercept lighting that otherwise would strike the wind turbine. In this case, the location factor decreases the equivalent collection area of the wind turbine by 75%. Objects smaller than the wind turbine within a distance $3H$ from it may still be instrumental in reducing the number of dangerous events due to flashes, providing that their collection area extends outside the wind turbine's own collection area.

Based on the above, the average number of direct lightning flashes N_D per year terminating directly on the WT blades can be estimated (4.9).

$$N_D = N_G C_D A_D 10^{-6}, \tag{4.9}$$

where N_G the lightning ground flash density.

As previously explained, upward lightning flashes tend to occur from tall structures of height in excess of 100 m, or from shorter structures situated on mountain tops; additionally, WT's rotating blades may initiate upward lightning.

Table 4.6 Location factor C_D.

Relative location	C_D
WT surrounded by higher objects	0.25
WT surrounded by objects of the same height or smaller	0.5
Isolated WT: no other objects in the vicinity	1
Isolated WT on a hilltop or a knoll	2
Offshore WTs	3 to 5

The number of upward lightning flashes N_U striking a WT can be determined with (4.10) [58].

$$N_U = 25 N_g^{0.8} e^{-\frac{E_{gc}-E_{g0}}{E_{g1}}}, \qquad (4.10)$$

where E_{g0} is the threshold electric field (2 kV/m for ordinary lightning-storm situations and a value of 3 kV/m for harsh lightning-storm locations) and E_{g1} is the shape parameter (2 kV/m and 6 kV/m for ordinary and harsh lightning-storm locations, respectively). E_{gc} is the critical ground electric field that depends on the height of the structure and can be empirically calculated as $E_{gc} = \frac{682.959}{H_{eff}} + 0.17776 \ln H_{eff}$ [58].

The number of failure per annum of a WT due to direct lightning strikes can be calculated as $N_{FWT} = (N_D + N_U) \int_{I_{MF}}^{\infty} f(I) dI$, where I_{MF} is the minimum lightning stroke currents causing a failure of a WT electrical components (e.g., 200 kA satisfies the design criteria for LPL I).

4.3.4 Assessment of the Average Annual Number of Dangerous Events N_L Due to Flashes Directly to and near Service Cables

Turbines may be connected via service underground cable of length L_c to an operation building of smaller height H_b. The collection areas of the connected building and of the buried cable should be calculated for the determination of the total collection area of the installation. Underground cables may be subjected to flashes striking them directly, and to flashes striking the earth near them.

The collection area A_L of flashes directly striking the underground cable is proportional to $\sqrt{\rho}$, where ρ is the soil resistivity, which is assumed in IEC

62305-2 to be 400 Ωm; the larger is the soil resistivity, the larger becomes the collection area. In this case, A_L is a rectangular surface of width equal to $2\sqrt{\rho}$, and length equal to the length of the cable L_L. If the length of the cable is unknown, 1,000 m is to be conservatively assumed. Lightning strikes within the area A_L may exceed the dielectric strength of the soil (3–5 kV/cm) and cause a discharge across soil and cable, which can cause fire and/or explosions.

The union of all single collection areas present on site, without any intersection, is the equivalent collection area to be used in the calculation of the annual number N_L of overvoltages of amplitude not less than 1 kV, possibly impacting the underground cable (Eq. 4.11).

$$N_L = N_G.A_L.C_I.C_E.C_T.10^{-6} \tag{4.11}$$

where N_G is the annual lightning ground flash density per square kilometer; C_I is the *installation* factor equal to 0.5, which takes into account a 50% reduction in the electromagnetic coupling of the underground cable with the lightning flash, if compared to an aerial installation, for which $C_I = 1$.

If the cable within the collection area A_L is at medium-voltage (exceeding 1kV), the step-up transformer, housed, for example, in the tower base, acts as a "buffer." The transformer, in fact, reduces the surges at the entry point into the tower, and this is taken into consideration by the *line type* factor C_T, which equals 0.2 ($C_T = 1$ for low-voltage power lines with no transformer). C_E is the *environmental* factor, which describes the objects that may surround the underground cable (see Table 4.7).

According to IEC 62305-2, the collection area A_1 of flashes to ground in the vicinity of the cable is proportional to the soil resistivity ρ, and is equal to 4,000 L_L. The annual number N_L of overvoltages of amplitude not lower than 1 kV occurring near the underground cable is still given by Eq. 4.11 by replacing A_L with A_1. Lightning flashes striking inside the area A_1 may induce transients and cause punctures to the cable insulation, which can cause short circuits.

Table 4.7 Environmental factor C_E.

Environment	C_E
Rural	1
Suburban	0.5
Urban	0.1
Urban with tall buildings (higher than 20 m)	0.01

4.3.5 Lightning Protection Zones

According to IEC 61400-24,[7] the WT may be separated into *Lightning Protection Zones* (LPZs) of different level of danger [35], which identify a specific electromagnetic environment, as follows:

1) **LPZ 0A**: Zone where the threat is due to the direct lightning flash and the full lightning electromagnetic field (e.g., this is typically the roof area of a structure without lightning protection). The internal systems may be subjected to full or partial lightning surge current. The nacelle and the blades fall within the LPZ 0A because they have to withstand both full lightning current and direct attachment; the d.c. side of the PV installation is usually considered in LPZ 0A without an external LPS.

2) **LPZ 0B**: Zone protected against direct lightning flashes but where the threat is the full lightning electromagnetic field (e.g., this is typically the sidewalls of a structure or a roof with structural lightning protection). The internal systems may be subjected to partial lightning surge currents.

3) **LPZ 1**: Zone where the surge current is limited by current sharing and by SPDs at the boundary (e.g., this is typically the area where services enter the structure or where the main switchboard is located within the nacelle). Spatial shielding may attenuate the lightning electromagnetic field.

4) **LPZ 2, ..., *n***: Zones where the surge current may be further limited by current sharing and by additional SPDs at the boundary (e.g., this may be a screened room, the main distribution panel, the control switchboard inside the nacelle, the PV inverter enclosures). Additional spatial shielding may be used to further attenuate the lightning electromagnetic field.

The higher the index number of a zone, the lower are the danger parameters.

The *Lightning Protection Level* (*LPL*), earlier defined as a number representing the lightning parameters that are not statistically exceeded by natural lightning, must be selected according to the established LPZ, so that to ensure the required protection. According to IEC 61400-24, the lightning protection of the WT must comply with the requirements of LPL I, which is the most stringent category, unless otherwise shown by risk analysis.

The *rolling sphere* is a method identified by IEC 61400-24 to determine the lightning protection zones, and the possible areas of the WT where a lightning is most likely to strike.

---

7 IEC 61400-24: "WT Systems, part 24, Lightning protection".

With this method, a sphere of radius corresponding to LPL I (i.e., 20 m) is rolled over the turbine, and all the resulting points of contact represent potential strike points. According to the IEC 61400-24, the rolling sphere method must not be used for the rotor blades.

This method identifies the LPZs 0A and LPZs 0B of the WT, which are respectively the zones that may be subjected to direct lightning strikes, and the zones that are inherently protected from direct lightning flashes (Figure 4.22).

Parts of the wind turbine within LPZs 0A need specific lightning rods for their protection, unlike parts within LPZ 0B. The identification of the different lightning protection zones of the WT does depends on its specific design.

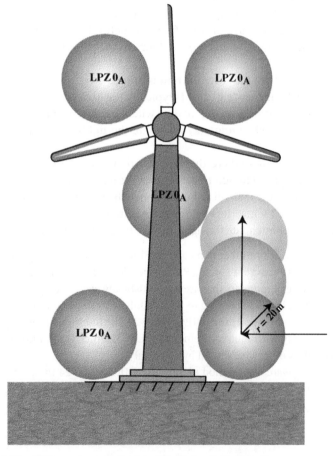

Figure 4.22 Rolling sphere method and Lightning Protection Zones.

4.4 High-Frequency Grounding Systems

Traditionally, at 50/60 Hz the ground electrode can be represented by a single resistor. For fast-varying currents, such as atmospheric discharge currents, the reactive components are no longer negligible.

In Figure 4.23, the equivalent lumped-parameter circuit of a vertical rod or a horizontal ground electrode of length l and radius a subjected to the lightning current $i(t)$ is shown: L_g is the inductance of the electrode, Z is the internal resistance of the metal electrode (usually negligible), and R_g and C_g are the resistance and the capacitive reactance to ground of the electrode, respectively.

Different sets of formulas for the parameters of the R-L-C circuit are available from the studies of Sunde [67], Rudenberg [68], and Tagg [69]. We can consider the following expressions that are based on the approximate method of the average potential [70] for $a \ll l$:

$$
R_G = \begin{cases} \dfrac{\rho_g}{2\pi l} W_{\text{ver}} \\[2ex] \dfrac{\rho_g}{\pi} W_{\text{hor}} \end{cases}, \quad
C_G = \begin{cases} \dfrac{2\pi\varepsilon_g l}{W_{\text{ver}}} \\[2ex] \dfrac{2\pi\varepsilon_g}{W_{\text{hor}}} \end{cases}, \quad
L_G = \begin{cases} \dfrac{\mu_g l}{2\pi} W_{\text{ver}} \\[2ex] \dfrac{\mu_g}{2\pi} W_{\text{hor}} \end{cases} \tag{4.12}
$$

with $W_{\text{ver}} = \ln\left(\dfrac{4l}{a}\right) - 1$ and $W_{\text{hor}} = \ln\left(\dfrac{2l}{\sqrt{2ad}}\right) - 1$, where d is the burial depth, ρ_g, ε_g and μ_g are the absolute electrical parameters of the soil.

As put in evidence by many studies, the behavior of grounding systems subject to high impulse current (e.g., lightning strikes) might be dramatically different from that at power frequency, and in some cases, the efficiency of the protection can be critically deteriorated. Two different physical processes are dominant in the dynamic behavior of grounding electrodes during transient discharge. First, the reactive component (i.e., inductive or capacitive) of ground electrodes may be more significant than their resistive part, which may affect their performance by increasing the grounding impedance during fast

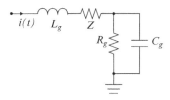

Figure 4.23 Equivalent circuit of a horizontal ground electrode.

rise-time lightning currents. Secondly, large currents dissipated into the surrounding earth can cause soil ionization around the electrode, making the electrode's impulse response strongly nonlinear. However, the nonlinear effects due to soil ionization might improve the electrode's performance by lowering its grounding impedance during high current discharge by virtually enlarging the dimensions of the electrode.

To better understand the main phenomena, in Figure 4.24 we show the theoretical case of the transient behavior of a long vertical rod buried into a low-resistivity soil under a negative first stroke lightning current $i(t)$ (median values taken from Table III in [61]). It is clear that the transient voltage rise $v(t)$ at the injection point into the rod in relation to the remote ground (often referred to as the *transient ground potential rise*) is very different from the mere ground potential rise $R_g i(t)$, where R_g is the power frequency ground resistance and $i(t)$ is the transient current flowing into the rod. In fact, even at frequencies above tens of kHz, the capacitive reactance $\dfrac{1}{\omega C_g}$ is still greater than R_g, and Z is negligible compared to the inductive reactance of the wire. Therefore, the equivalent lumped-parameter circuit of the ground electrode in impulse conditions can be simplified to only include R_g and L_g, which create an inductive ground impedance to the lightning pulse.

The frequency behavior of the normalized harmonic impedance $\dfrac{Z_g(\omega)}{R_g}$ is shown in Figure 4.25. At the inception of the stroke, a high frequency is associated with the fast-rising front of the lightning current during; therefore, a high inductive reactance adds to the ground resistance of the electrode. On the

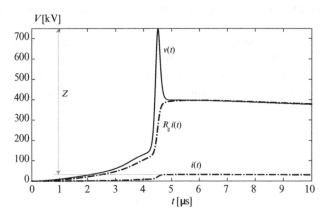

Figure 4.24 Transient potential rise at the top of a vertical rod under a first stroke lightning current ($\rho_g = 50\ \Omega\text{m}$, $\varepsilon_g = 10\varepsilon_0$ F/m, $l = 4$ m, $a = 12.5$ mm).

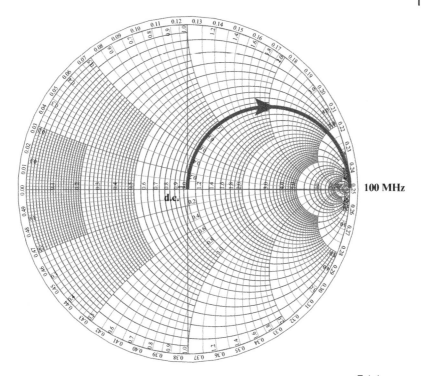

Figure 4.25 Frequency behavior of the normalized harmonic impedance $\dfrac{Z_g(\omega)}{R_g}$ for the vertical rod of the previous figure.

other hand, a lower frequency exists in the wave-tail of the decaying lightning current; thus, the inductive reactance drops. The example of the vertical rod shows that the high-frequency inductive behavior of the grounding electrode might result in large peaks of transient potential at the feed point in cases when the current pulses have enough high-frequency content. However, usually after a few microseconds, the transient behavior finishes, and the ground impedance $z_g(t)$ settles to the value of the d.c. resistance R_g.

As earlier mentioned, when a high magnitude lightning current flows into the earth, the electromagnetic field around the electrode may exceed the dielectric strength of the soil (typically ranging between 100 and 500 kV m^{-1}), which may break down, and soil ionization occurs. This dramatically reduces the ground impedance $z(t)$, thereby improving the performance of the electrode. After the decay of the lightning current, and of the associated electromagnetic field, the soil ionization stops, and the soil recovers its original resistivity (Figure 4.26).

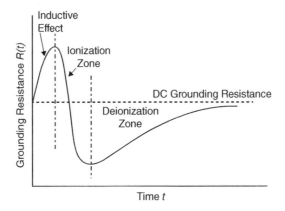

Figure 4.26 Grounding impedance as a function of time.

As defined in [71], the transient potential $v(t)$ can be expressed as $v(t) = R_g i(t) + x(t)$, where the first term approximates the linear resistive ground potential rise, while the second term is related to the frequency-dependent phenomena and is an approximation of the reactive component of the transient potential.

The transient characteristic of a grounding systems mostly depends upon parameters related to the ground electrodes' geometry, the soil's electrical properties, and the lightning current waveform properties, i.e., current intensity and current pulse front time. With reference to Figure 4.24, we can define several quantities to characterize the dynamic behavior of grounding systems. We can define the *transient grounding impedance* $z_g(t) = \dfrac{v(t)}{i(t)}$, which varies with time under impulse currents, exhibiting a rapid variation during the initial surge period, after which it converges to the stationary condition characterized by a nearly constant value that can be approximated by the low-frequency grounding resistance [72]. The transient grounding impedance is mainly used to distinguish two time periods of the transient response: the initial surge, or fast transient period, and the consequent stationary, or slow transient, period [71].

The *impulse impedance* $Z_g = \dfrac{V_{peak}}{I_{peak}}$ is defined as the ratio of the peak value of the potential rise V_{peak} to the peak value of the injected current I_{peak} through the electrode. The impulse grounding impedance does not have any physical meaning, since the two peak values usually take place at different times. It is, however, a very useful parameter in lightning protection because if the lightning current is known (e.g., chosen according to applicable Standards), then we can estimate the maximum potential of the grounding electrode.

To compare the performances of grounding electrode under surge conditions and at power frequency, Z_g may be related to the power frequency grounding resistance R_g through the dimensionless *impulse coefficient* $A_g = \dfrac{Z_g}{R_g}$ [73]. Values of A_g larger than one indicate an impaired surge performance compared to the low-frequency grounding performance; values of A_g less than one indicate a superior surge performance of the electrode.

The *harmonic impedance* $Z_g(\omega) = \dfrac{V(\omega)}{I(\omega)}$ is a useful parameter in transient analysis [74], which is defined as the ratio of the steady-state harmonic electric potential $V(\omega)$ at the feed point, in reference to the remote earth, to the injected current $I(\omega)$, with the frequency ranging between d.c. to the highest frequency present in the lightning current.

4.4.1 Arrangement of Ground Electrodes

Ground electrodes should safely conduct the lightning current into the ground, equalize the potential between the connected down-conductors, and limit step and touch potentials [75]. Critical values of step and touch voltages are based on the maximum energy tolerated by a person's body in lightning conditions, which is assumed to be 20 J.

The IEC 62305-3[8] standard describes two basic types of ground-electrode arrangements for lightning protection purposes: type A and type B.

Type A consists of horizontal or vertical ground electrodes, each connected to a down-conductor. A ring ground electrode interconnecting the down-conductors is also classified as type A, if it is buried for less than 80% of its length. In a type A electrode arrangement, the minimum number of ground electrodes is one for each down-conductor, with a minimum of two for the whole lightning protection system. According to IEC 61400-24, type A arrangements must not be used for wind turbines but may be used for the buildings used in connection to the turbine, such as structures with measurement equipment, office sheds, etc.

The type B arrangement, to be used with WTs, includes a ring conductor external to the structure buried for at least 80 % of its total length, interconnecting the down-conductors; or a foundation concrete-encased electrode forming a closed loop. The requirement of the closed-loop connection for type B electrodes, even though 20 % may not be in contact with the soil, guarantees an equal division of the lightning current through the electrode, and therefore a

---

8 IEC 62305-3: 2010 "*Protection against lightning* –. *Part 3: Physical damage to structures and life hazard*".

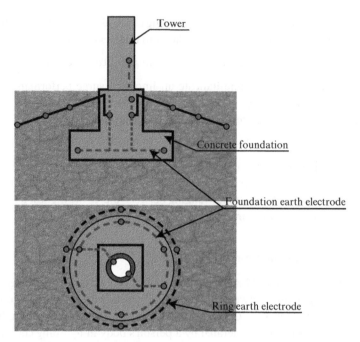

Figure 4.27 Ground-termination system of a WT.

uniform dissipation into the soil. Both ring and the re-bars must be connected to the conductive tower (Figure 4.27).

IEC 62305-3 requires for both ring and foundation ground electrodes that the mean radius r_e of the equivalent circular area enclosed by the electrode be not less than the value l_1 that is provided in the standard according to LPS classes (i.e., I, II, III, and IV) and soil resistivity.

The IEC 61400-24 requires that, unless otherwise shown by risk analysis, wind turbines should always be protected against lightning as per the most stringent requirements of class LPS I.

According to IEC 62305-3, for soil resistivities, less than 500 Ωm, the length l_1 has a constant value of 5 m, whereas for soil resistivities ranging between 500 Ωm and 3,000 Ωm, the length l_1 linearly increases up to 80 m.

If r_e is less than the prescribed value l_1, a combination of horizontal, vertical or inclined electrodes should be added and connected to the electrode. The individual lengths l_r (for the horizontal electrodes) and l_v (for the vertical electrodes) must satisfy equations Eqs. (4.13) and (4.14).

$$l_r = l_1 - r_e \tag{4.13}$$

$$l_v = \frac{l_1 - r_e}{2} \qquad (4.14)$$

The above equations show how the standard deems more effective vertical electrodes, which may be shorter. The additional vertical and horizonal electrodes should be connected to the ring ground electrode at the same points of connection as the down-conductors.

4.4.2 Effective Length of a Ground Electrode

Due to the self-inductance of ground wires, only the sections of the grounding system closest to the lightning injection point can effectively disperse the current into the earth. The lightning ground impedance decreases with the length of the electrode into the earth but saturates beyond a specific value.

The *effective length* l_e of an electrode is practically defined as the minimum length of a ground wire taken from the current injection point to a point that allows the dispersion into the soil of at least 95% of the lightning current. The effective length identifies the length of the ground electrode beyond which the ground impedance does not significantly decrease; therefore, burying longer electrodes is neither effective nor economically justifiable. A more theoretical definition can be found in [76–78]: *the effective length is the maximum length of a grounding wire over which the impulse impedance Z_g value versus a unique lightning current that flows into the grounding wire, ceases to decrease even if the grounding wire is lengthened.* Grcev [1, 71, 73] mathematically defines the effective length as the *maximum grounding electrode length for which the impulse coefficient A_g is equal to one*. In fact, the impulse impedance Z_g of ground electrodes with smaller length is equal to the low-frequency ground resistance R_g, that is, $A_g = 1$ (resistive behavior). The impulse impedance decreases with the increase in the electrode length, but at a certain length, it becomes constant, while the low-frequency resistance continues to decrease, resulting in impulse coefficient larger than one (reactive behavior). He et al. [79] define the effective length as the maximum wire length fulfilling the equation $\frac{dZ_g}{dl} \le \tan\vartheta$, where ϑ is assumed to be 5°.

In addition to the above definitions, another school of thought [80] defines the effective length as *the maximum wire length at which, when comparing (1) the value of the rise in inductive electrical potential of the wave front at the injection point and (2) the value of the rise in electrical potential when the electrical potential of the injection point rises again as the lightning surge in voltage along the grounding wire, having traveled to the tip of the grounding wire, becomes a positive reflective wave and travels back toward the injection point, the latter (2) does not exceed the former (1)* (for a discussion on these definitions the interested reader is referred to [79]).

Experiments show that l_e increases with the soil resistivity ρ_g, as the current preferably tends to flow through the electrode itself rather than through the low-conductivity volume of the soil near the point of injection. The effective length also increases with the wave front time τ (in μs) of the impulse current but decreases with the magnitude of lightning current I_{peak} (in kA). Soil permittivity has practically no impact on the effective length. A simple formula for effective length of horizontal electrodes that summarizes the above concepts has been derived through least squares curve fitting methods: $l_e = \dfrac{6.528\left(\rho_g \tau\right)}{I_{peak}^{0.097}}$.

The effective length of horizontal or vertical buried ground wires connected to the re-bars of the structural foundation of the wind turbine can be calculated as $l_e = k\sqrt{\rho_g \tau}$, where $1.30 \le k \le 1.43$ for horizontal wires and $k \cong 0.71$ for vertical wires [81].

The reactive behavior of short electrodes (i.e., length less than 3 m) is usually capacitive, and may be disregarded, except in very resistive soil and for fast rise-time pulses, for which rapid potential variations are filtered out during the rising portion of the pulse. Conversely, the behavior of longer electrodes (i.e., greater than 30 m) is more complex. The reactive behavior increases with fast rise-time pulses: it is dominantly inductive in conductive soil and becomes dominantly capacitive in resistive soil. In resistive soils, the capacitive effects improve the grounding performance by effectively filtering out rapidly varying voltages during surges.

Given a soil resistivity, the optimum performance of WT grounding systems in impulse conditions is reached by designing ground electrodes with length not exceeding l_e, and increasing the number of conductors, for example, arranging them in a star configuration, with the down-conductor connected at the center.

4.4.3 Frequency-dependent Soil and Ionization

When performing accurate studies of the transient behavior of grounding systems, frequency-dependent behavior and soil ionization are important factors that must be accounted for. A complete grounding system model that incorporates both effects becomes very complex, i.e., a nonlinear time-variant frequency-dependent system. The major difficulty lies in considering both effects at the same time. In fact, although the frequency-domain approach is well suited for the analysis of grounding systems by considering the frequency-dependence parameters, only time-domain simulations can incorporate soil ionization nonlinearities. It is, therefore, rather challenging to consider the frequency and current dependences simultaneously.

The electrical parameters of the soil depend on a combination of its different atomic properties, as well as on various mechanisms of conduction and losses occurring at different frequency intervals [82]. Several models are available in literature for the evaluation of the ground conductivity $\sigma_g(f)$ and $\varepsilon_g(f)$ as a function of the frequency (e.g., Scott, Smith and Longmire, Messier, Visacro and Portela, Portela, Visacro and Alipio); a useful discussion about different soil models and their influence on the GPR can be found in [83]. According to these models, a complex effective permittivity (or conductivity) can be calculated as $\bar{\varepsilon}_g = \varepsilon_g - j\dfrac{\sigma_g}{\omega}$ (or $\bar{\sigma}_g = \sigma_g + j\omega\varepsilon_g$). Particular attention must be paid to ensure causality: the Kramers–Kronig relationship must be satisfied by both real and imaginary parts of the effective permittivity. It has been observed that the general trend of the frequency dependence of the electrical parameters of the soil leads to a decrease of both the resistivity and permittivity, compared to low-frequency values. Recent studies have revealed that the soil parameters variation with the frequency results in a decrease in the potential rise of the grounding electrodes under direct lightning strikes, with respect to the case where the parameters are assumed constant. However, the effect of the soil parameters variation with the frequency on the GPR developed by indirect lightning was found negligible [84].

When a current is injected into the soil, a spatial electric field distribution **E** is generated in accordance to the Ohm's law $\mathbf{E} = \rho\mathbf{J}$, where **J** is the current density vector. Generally, this relation is time-dependent. Experimental and theoretical studies have shown that, when the surge current density **J** leaking into the soil by the different parts of the earthing system increases, the electric field **E** on the lateral surface of the electrodes can overcome the electrical strength, and the non-linear phenomena due to the soil ionization (or soil breakdown) take place [85]. The soil ionization region starts at the electrode surface, where the current density has its highest value. This region extends up to a distance where the current density decreases to a value that makes the electric field lower than the critical breakdown value for the soil. When the surge current into the soil begins to decrease, the electric field in the ionized region also begins to decrease, and the deionization process takes place, which restores the stationary steady soil characteristics.

At a micro-structural level, most soils basically consist of non-conducting particles coated with water in which some salt is dissolved, with air filling the voids between the particles. The average size of the air voids within the soil will depend on the frequency distribution of the size of the particles. For instance, a soil consisting of a very fine dust-like particles will have smaller void sizes, while a sandy soil with coarse particles will have larger void sizes. The shape of the voids is usually highly irregular, especially if the surrounding particles have sharp edges.

When the electric field in the void becomes large enough, the air ionization process may begin and the current, that until that moment flows through the water paths within the soil, is now mainly conducted by the ionized air. Due to the irregular shape of the voids and the effect of the relatively large dielectric constant of the soil, the critical value of the electric field E_c can be much smaller than the breakdown field of an equivalent air gap, and it ranges between 1 and 3 kV/cm [86, 87].

In technical literature, two sets of models are mainly employed to describe the soil ionization and deionization processes.

The first school of thought is the *time-variable soil resistivity approach* [86, 88–91], which considers a time variable soil resistivity in the region surrounding the earth electrode. The time-variable resistivity is a non-linear function of the electric field, and during the ionization process, when $E > E_c$, it can be expressed as $\rho = \rho_0 \exp\left(-\dfrac{t}{\tau_1}\right)$, where τ_1 is the ionization time constant of the soil. During the de-ionization process, the resistivity can be expressed as $\rho = \rho_{min} + (\rho_0 - \rho_{min})\left(1 - e^{-\frac{t}{\tau_2}}\right)\left(1 - \dfrac{E}{E_c}\right)^2$, where ρ_{min} is the minimal values reached by the soil resistivity during the ionization process, and τ_2 is the deionization time constant of the soil.

The second school of thought is the *time-variable electrodes geometry approach* [92–95], which is the most commonly used. This approach considers a given electrode embedded in an ionized soil as an electrode of modified transversal dimensions buried into a non-ionized soil. This approach considers the soil resistivity unchanged, and the non-linear behavior is described by the effect of the equivalent electrode geometry on the current flowing into the soil. For each value of the ground current, the effective radius of the electrode is obtained by assuming that the electric field may not exceed the critical value: hence, the radius of each conductor segment is chosen to be variable as a function of the electric field $E(t)$. With the time-variable electrodes geometry approach, the ionized region is similar to the conductor, and the electric field in this region is assumed to be roughly zero, as if the ionized region were short-circuited to the electrode.

Since soil ionization usually improves the ground-electrode performance, not considering it may be considered as a choice on the "safe" side. Interesting studies for WTs are available in [96–99] and for photovoltaic installations in [100]: it is generally observed that the ionization phenomenon reduces the total grounding resistance and the ground-potential rise; this becomes more evident when the soil resistivity and the electrode dimensions (i.e., the apparent grounding resistance) are elevated.

References

1 Rakov, V.A. and Rachidi, F. (2009). Overview of recent progress in lightning research and lightning protection. *IEEE Transactions on Electromagnetic Compatibility* 51 (3, PART 1); 428–442.

2 Uman, M.A. and Krider, E.P. (1982). A review of natural lightning: Experimental data and modeling. *IEEE Transactions on Electromagnetic Compatibility* EMC-24 (2): 79–112.

3 Bazelyan, E.M. and Raizer, Y.P. (2000). *Lightning Physics and Lightning Protection*. Bristol, UK: IOP Publishing.

4 Cooray, G.V., *The Lightning Flash*. London, UK: Institution of Electrical Engineers, 2002.

5 Rakov, V.A. and Uman, M.A. (2003). *Lightning: Physics and Effects*. Cambridge, UK: Cambridge University Press.

6 Rachidi, F. and Tkachenko, S.V. (2008). *Electromagnetic Field Interaction with Transmission Lines: From Classical Theory to HF Radiation Effects*. Southampton, UK: WIT Press.

7 Uman, M.A. (2008). *The Art and Science of Lightning Protection*. Cambridge, UK: Cambridge University Press.

8 Laroche, P., Schumann, U., and Betz, H.D. (2009). *Lightning: Principles, Instruments and Applications*. New York: Springer-Verlag.

9 Rakov, V.A. (2016). *Fundamentals of Lightning*. Cambridge, UK: Cambridge University Press.

10 CIGRE WG C4.407. (2013). Lightning Parameters for Engineering Applications.

11 Diendorfer, G., Pichler, H., and Mair, M. (August 2009). Some parameters of negative upward-initiated lightning to the Gaisberg tower (2000–2007). *IEEE Transactions on Electromagnetic Compatibility* 51 (3): 443–452.

12 Betz, H.D., Schumann, U., and Laroche, P. (2009). *Lightning: Principles, Instruments and Applications*. The Netherlands: Springer.

13 Wang, D., Takagi, N., Watanabe, T., Sakurano, H., and Hashimoto, M. (January 2008). Observed characteristics of upward leaders that are initiated from a windmill and its lightning protection tower. *Geophysical Research Letters* 35 (2): L02803.

14 Rachidi, F., et al. (2008). A review of current issues in lightning protection of new-generation wind-turbine blades. *IEEE Transactions on Industrial Electronics* 55 (6): 2489–2496.

15 Montanyà, J., Van Der Velde, O., and Williams, E.R. (February 2014). Lightning discharges produced by wind turbines. *Journal of Geophysical Research* 119 (3): 1455–1462.

16 Lu, W., Wang, D., Zhang, Y., and Takagi, N. (March 2009). Two associated upward lightning flashes that produced opposite polarity electric field changes. *Geophysical Research Letters* 36 (5).

17 Diendorfer, G. (2015). On the risk of upward lightning initiated from wind turbines. *2015 IEEE 15th International Conference on Environment and Electrical Engineering (EEEIC)*, 872–876.

18 Wen, X., et al. (2016). Effect of wind turbine blade rotation on triggering lightning: An experimental study. *Energies* 9 (12):

19 Peesapati, V., Cotton, I., Sorensen, T., Krogh, T., and Kokkinos, N. (January 2011). Lightning protection of wind turbines - a comparison of measured data with required protection levels. *IET Renewable Power Generation* 5 (1): 48–57.

20 Heidler, F.H. (2019). Review and extension of the TCS model to consider the current reflections at ground and at the upper end of the lightning channel. *IEEE Transactions on Electromagnetic Compatibility* 61 (3): 644–652.

21 Szedenik, N. (2001). Rolling sphere - Method or theory? *Journal of Electrostatics* 51: 345–350.

22 Singhasathein, A., Rungseevijitprapa, W., and Pruksanubal, A. (2019). The modern mathematical equation for the lightning protection angle. *2018 15th International Conference on Electrical Engineering/Electronics, Computer, Telecommunications and Information Technology (ECTI-CON 2018)*, 249–252.

23 De Araújo, B.R. (2017). Mathematical modeling for analysis and design of LPS: Angle method. *2017 International Symposium on Lightning Protection (XIV SIPDA)*, 42–48.

24 Hasse, P., Wiesinger, J., and Zischank, W. (2006). *Handbuch Fur Blitzschutz Und Erdung*, Vol. 5. Munchen, Germany: Auflage.

25 Arevalo, L. and Cooray, V. (2010). 'The mesh method' in lightning protection standards - Revisited. *Journal of Electrostatics* 68 (4): 311–314.

26 Formisano, A., Petrarca, C., Hernández, J.C., and Munōz-Rodríguez, F.J. (2019). Assessment of induced voltages in common and differential-mode for a PV module due to nearby lightning strikes. *IET Renewable Power Generation* 13 (8): 1369–1378.

27 Qi, R., Du, Y., and Chen, M. (November 2019). Lightning-generated transients in buildings with an efficient PEEC method. *IEEE Transactions on Magnetics* 55 (11): 1–5.

28 Ahmad, N.I., et al. (2018). Lightning protection on photovoltaic systems: A review on current and recommended practices. *Renewable and Sustainable Energy Reviews* 82 (March 2017): 1611–1619.

29 Ittarat, S., Hiranvarodom, S., and Plangklang, B. (2013). A computer program for evaluating the risk of lightning impact and for designing the installation of lightning rod protection for photovoltaic system. *Energy Procedia* 34 (March): 318–325.

30 Zhang, Y., Chen, H., and Du, Y. (2020). Considerations of photovoltaic system structure design for effective lightning protection. *IEEE Transactions on Electromagnetic Compatibility* 62 (4): 1333–1341.

31 Kokkinos, N., Christofides, N., and Charalambous, C. (2012). Lightning protection practice for large-extended photovoltaic installations. *2012 International Conference on Lightning Protection (ICLP)*, 1–5.

32 Van Brunt, R.J., Nelson, T.L., and Stricklett, K.L. (2000). Early streamer emission lightning protection systems: An overview. *IEEE Electrical Insulation Magazine* 16 (1): 5–24.

33 Carpenter, R.B. and Drabkin, M.M. (1998). Protection against direct lightning strokes by Charge Transfer System. *1998 IEEE EMC Symposium. International Symposium on Electromagnetic Compatibility. Symposium Record (Cat. No.98CH36253)*, vol. 2, 1094–1097.

34 Tu, Y., Zhang, C., Hu, J., Wang, S., Sun, W., and Li, H. (2013). Research on lightning overvoltages of solar arrays in a rooftop photovoltaic power system. *Electric Power Systems Research* 94: 10–15.

35 Hernández, J.C., Vidal, P.G., and Jurado, F. (2008). Lightning and surge protection in photovoltaic installations. *IEEE Transactions on Power Delivery* 23 (4): 1961–1971.

36 Naxakis, I., Christodoulou, C., Perraki, V., and Pyrgioti, E. (2017). Degradation effects on single crystalline silicon photovoltaic modules subjected to high impulse-voltages. *IET Science, Measurement & Technology* 11 (5): 563–570.

37 Wang, Y., Zhang, X., and Tao, S. (2019). Modeling of lightning transients in photovoltaic bracket systems. *IEEE Access* 7: 12262–12271.

38 Charalambous, C.A., Kokkinos, N.D., and Christofides, N. (April 2014). External lightning protection and grounding in large-scale photovoltaic applications. *IEEE Transactions on Electromagnetic Compatibility* 56 (2): 427–434.

39 Rachidi, F. (2012). A review of field-to-transmission line coupling models with special emphasis to lightning-induced voltages on overhead lines. *IEEE Transactions on Electromagnetic Compatibility* 54 (4): 898–911.

40 Denno, K. (1984). Computation of electromagnetic lightning response using moments method. *IEEE Transactions on Magnetics* 20 (5): 1953–1955.

41 Laudani, A.A.M., Carloni, L., Thomsen, O.T., Lewin, P., and Golosnoy, I.O. (2020). Efficient method for the computation of lightning current distributions in wind turbine blades using the Fourier transform and the finite element method. *IET Science, Measurement & Technology* 14 (7): 786–799.

42 Baba, Y. and Rakov, V.A. (2014). Applications of the FDTD method to lightning electromagnetic pulse and surge simulations. *IEEE Transactions on Electromagnetic Compatibility* 56 (6): 1506–1521.

43 Ehrhardt, A. and Beier, S. (2014). Spark gaps for DC applications. *ICEC 2014; The 27th International Conference on Electrical Contacts*, 1–6.

44 He, J., Yuan, Z., Wang, S., Hu, J., Chen, S., and Zeng, R. (2010). Effective protection distances of low-voltage SPD with different voltage protection levels. *IEEE Transactions on Power Delivery* 25 (1): 187–195.

45 Jinliang, H., Yuan, Z., Jing, X., Chen, S., Zou, J., and Zeng, R. (2005). Evaluation of the effective protection distance of low-voltage SPD to equipment. *IEEE Transactions on Power Delivery* 20 (1): 123–130.

46 Rodrigues, R.B., Mendes, V.M.F., and Catalão, J.P.S. (2012). Analysis of transient phenomena due to a direct lightning strike on a wind energy system. *Energies* 5 (7): 2545–2558.

47 Napolitano, F., *et al.* (2011). Models of wind-turbine main-shaft bearings for the development of specific lightning protection systems. *IEEE Transactions on Electromagnetic Compatibility* 53 (1): 99–107.

48 Peesapati, V., Cotton, I., Sorensen, T., Krogh, T., and Kokkinos, N. (2011). Lightning protection of wind turbines - a comparison of measured data with required protection levels. *IET Renewable Power Generation* 5 (1): 48–57.

49 Sarajčev, P. and Goić, R. (April 2011). A review of current issues in state-of-art of wind farm overvoltage protection. *Energies* 4 (4): 644–668.

50 Cotton, I., Jenkins, N., and Pandiaraj, K. (2001). Lightning protection for wind turbine blades and bearings. *Wind Energy* 4: 23–37.

51 McNiff, B. (2020). Wind turbine lightning protection project. *NREL Subcontract. Rep. No. SR-500-31115'*.

52 Madsen, S.F., Bertelsen, K., Krogh, T.H., Erichsen, H.V., Hansen, A.N., and Lønbæk, K.B. (2010). Proposal of new zoning concept considering lightning protection of wind turbine blades. *2010 30th International Conference on Lightning Protection (ICLP)*, 1–7.

53 Yokoyama, S. (2013). Lightning protection of wind turbine blades. *Electric Power Systems Research* 94: 3–9.

54 Shulzhenko, E., Krapp, M., Rock, M., Thern, S., and Birkl, J. (2017). Investigation of lightning parameters occurring on offshore wind farms. *2017 International Symposium on Lightning Protection (XIV SIPDA)*, 169–175.

55 Byrne, A. and Malkin, M. Field performance assessment of wind turbine lightning protection systems uncertainty due to upward lightning and other factors.

56 Borghetti, A., Nucci, C.A., and Paolone, M. (2004). Estimation of the statistical distributions of lightning current parameters at ground level from the data recorded by instrumented towers. *IEEE Transactions on Power Delivery* 19 (3): 1400–1409.

57 Eriksson, A.J. (1987). The incidence of lightning strikes to power lines. *IEEE Power Engineering Review* PER-7 (7): 66–67.

58 Rizk, F.A.M. (1994). Modeling of lightning incidence to tall structures. I. Theory. *IEEE Transactions on Power Delivery* 9 (1): 162–171.

59 Rizk, F.A.M. (1994). Modeling of lightning incidence to tall structures. II. Application. *IEEE Transactions on Power Delivery* 9 (1): 172–193.

60 Petrov Nikolai, N. and Nikolay. (January 2000). Determination of attractive area and collection volume of earthed structures. *Proc. ICLP'2000*.

61 Gamerota, W.R., Elismé, J.O., Uman, M.A., and Rakov, V.A. (2012). Current waveforms for lightning simulation. *IEEE Transactions on Electromagnetic Compatibility* 54 (4): 880–888.

62 Chowdhuri, P., et al. (2005). Parameters of lightning strokes: A review. *IEEE Transactions on Power Delivery* 20 (1): 346–358.

63 Petrov, N.I. and D'Alessandro, F. (December 2002). Assessment of protection system positioning and models using observations of lightning strikes to structures. *Proceedings. Mathematical, Physical, And Engineering Sciences / The Royal Society* 458 (2019): 723–742.

64 Shindo, T. (2018). Lightning striking characteristics to tall structures. *IEEJ Transactions on Electrical and Electronic Engineering* 13 (7): 938–947.

65 Zhou, H., Theethayi, N., Diendorfer, G., Thottappillil, R., and Rakov, V.A. (2010). On estimation of the effective height of towers on mountaintops in lightning incidence studies. *Journal of Electrostatics* 68 (5): 415–418.

66 Shindo, T. (2012). A calculation method of effective height of structures in lightning studies. *IEEJ Transactions on Electrical and Electronic Engineering* 132 (3): 292–293.

67 Sunde, E.D. (1949). *Earth Conduction Effects in Transmission Systems*. D. Van Nostrand Company.

68 Rüdenberg, R. (01/01/1968). *Electrical Shock Waves in Power Systems: Traveling Waves in Lumped and Distributed Circuit Elements*. Cambridge (MA): Harvard University Press.

69 Tagg, G.F. (1964). *Earth Resistances*. New York: Pitman Publishing Corporation.

70 Dwight, H.B. (1936). Calculation of resistances to ground. *Transactions of the American Institute of Electrical Engineers* 55 (12): 1319–1328.

71 Grcev, L. (2009). Time- and frequency-dependent lightning surge characteristics of grounding electrodes. *IEEE Transactions on Power Delivery* 24 (4): 2186–2196.

72 Velazquez, R. and Mukhedkar, D. (1984). Analytical modelling of grounding electrodes transient behavior. *IEEE Transactions on Power Apparatus and Systems* PAS-103 (6): 1314–1322.

73 Grcev, L. (2009). Impulse efficiency of ground electrodes. *IEEE Transactions on Power Delivery* 24 (1): 441–451.

74 Roubertou, D., Fontaine, J., Plumey, J.P., and Zeddam, A. (1984). Harmonic input impedance of earth connections. *1984 International Symposium on Electromagnetic Compatibility*, 1–4.

75 Araneo, R. and Celozzi, S. (2016). Transient behavior of wind towers grounding systems under lightning strikes. *International Journal of Energy and Environmental Engineering* 7 (2).

76 Gupta, B.R. and Thapar, B. (November 1980). Impulse impedance of grounding grids. *IEEE Transactions on Power Apparatus and Systems* PAS-99 (6): 2357–2362.

77 Jinliang, H., *et al.* (April 2005). Effective length of counterpoise wire under lightning current. *IEEE Transactions on Power Delivery* 20 (2): 1585–1591.

78 Pariyavong, P. and Rungseevijitprapa, W. (2008). Effective length consideration with the minimize per-unit length impulse impedance method. *IEEJ Transactions on Power and Energy* 128 (6): 871–878.

79 Yamamoto, K., Sumi, S., Sekioka, S., and He, J. (November 2015). Derivations of effective length formula of vertical grounding rods and horizontal grounding electrodes based on physical phenomena of lightning surge propagations. *IEEE Transactions on Industry Applications* 51 (6): 4934–4942.

80 Sekioka, S. and Funabashi, T. (2009). A study on effective length of long grounding conductor in windfarm. *IEEJ Transactions on Power and Energy* 129 (5): 675–681.

81 Miyamoto, S., Baba, Y., Nagaoka, N., and Yamamoto, K. (2017). Effective length of vertical grounding wires connected to wind turbine foundation. *Journal of International Council on Electrical Engineering* 7 (1): 89–95.

82 Akbari, M., *et al.* (2013). Evaluation of lightning electromagnetic fields and their induced voltages on overhead lines considering the frequency dependence of soil electrical parameters. *IEEE Transactions on Electromagnetic Compatibility* 55 (6): 1210–1219.

83 Cavka, D., Mora, N., and Rachidi, F. (February 2014). A comparison of frequency-dependent soil models: application to the analysis of grounding systems. *IEEE Transactions on Electromagnetic Compatibility* 56 (1): 177–187.

84 Nazari, M., Moini, R., Fortin, S., Dawalibi, F.P., and Rachidi, F. (2020). Impact of frequency-dependent soil models on grounding system performance for direct and indirect lightning strikes.in *IEEE Transactions on Electromagnetic Compatibility*, 1–11, vol. 63, no. 1, pp. 134–144, Feb. 2021, doi: 10.1109/TEMC.2020.2986646.

85 He, J. and Zhang, B. (November 2015). Progress in lightning impulse characteristics of grounding electrodes with soil ionization. *IEEE Transactions on Industry Applications* 51 (6): 4924–4933.

86 Mousa, A.M. (1994). The soil ionization gradient associated with discharge of high currents into concentrated electrodes. *IEEE Transactions on Power Delivery* 9 (3): 1669–1677.

87 Bellaschi, P.L. (1941). Impulse and 60-cycle characteristics of driven grounds. *Electrical Engineering* 60 (3): 123–127.

88 Liew, A.C. and Darveniza, M. (1974). Dynamic model of impulse characteristics of concentrated earths. *Proceedings of the Institution of Electrical Engineers* 121 (2): 123–135(12).

89 Oettle, E.E. (1988). A new general estimation curve for predicting the impulse impedance of concentrated earth electrodes. *IEEE Transactions on Power Delivery* 3 (4): 2020–2029.

90 Almeida, M.E. and Correia De Barros, M.T. (1996). Accurate modelling of rod driven tower footing. *IEEE Transactions on Power Delivery* 11 (3): 1606–1609.

91 Ala, G., Francomano, E., Toscano, E., and Viola, F. (2004). Finite difference time domain simulation of soil ionization in grounding systems under lightning surge conditions. *Applied Numerical Analysis & Computational Mathematics* 1 (1): 90–103.

92 Cidras, J., Otero, A.F., and Garrido, C. (2000). Nodal frequency analysis of grounding systems considering the soil ionization effect. *IEEE Transactions on Power Delivery* 15 (1): 103–107.

93 Geri, A. (1999). Behaviour of grounding systems excited by high impulse currents: The model and its validation. *IEEE Transactions on Power Delivery* 14 (3): 1008–1017.

94 Zhang, B., *et al.* (2005). Numerical analysis of transient performance of grounding systems considering soil ionization by coupling moment method with circuit theory. *IEEE Transactions on Magnetics* 41 (5): 1440–1443.

95 Moradi, M. (2020). Analysis of transient performance of grounding system considering frequency-dependent soil parameters and ionization. *IEEE Transactions on Electromagnetic Compatibility* 62 (3): 785–797.

96 Prousalidis, J.M., Philippakou, M.P., Hatziargyriou, N.D., and Papadias, B.C. (2000). The effects of ionization in wind turbine grounding modeling. *2000 10th Mediterranean Electrotechnical Conference. Information Technology and Electrotechnology for the Mediterranean Countries. Proceedings. MeleCon 2000 (Cat. No.00CH37099)*, vol. 3, 940–943.

97 Pyrgioti, E. and Bokogiannis, V. (2011). Lightning impulse performance of a wind generator grounding grid considering soil ionization. *2011 7th Asia-Pacific International Conference on Lightning*, 103–107.

98 Zhang, T., Zhang, Y., and Tan, X. (2014). Study on overvoltage of signal line in wind turbine by lightning strike. *2014 International Conference on Lightning Protection (ICLP)*, 1516–1519.

99 Markovski, B., Grcev, L., and Arnautovski-Toseva, V. (2012). Transient characteristics of wind turbine grounding. *2012 International Conference on Lightning Protection (ICLP)*, 1–6.

100 Naxakis, I., Mihos, G., Pastromas, S., and Pyrgioti, E. (2018). Examining the operation of the grounding system of a PV installation. *2018 IEEE International Conference on High Voltage Engineering and Application (ICHVE)*, 1–4.

5

Renewable Energy System Protection and Coordination

CONTENTS

Now is the winter of our discontent
made glorious summer by this son of York.

Shakespeare, Richard III

5.1 Introduction

The use of renewable energies to supply electricity has grown dramatically in the last years, which has produced a decrease in the cost of systems. Today, several utility-interactive systems, especially wind and photovoltaic power, are at grid parity in several countries, that is, that the *levelized cost* of generated renewable energy[1] (LCOE) is less than or equal to the cost of the energy

---

1 The levelized cost of electricity is a measure of the average net present cost of electricity generation for a generating plant over its lifetime.

Electrical Safety Engineering of Renewable Energy Systems. First Edition. Rodolfo Araneo and Massimo Mitolo.
© 2022 by The Institute of Electrical and Electronics Engineers, Inc. Published 2022 by John Wiley & Sons, Inc.

purchased from the grid. We also see an increasing concern over global climate change, which has resulted in a reduction of CO_2 emissions thanks to the replacement of fossil fuel-generated electricity with renewable energy systems. All these factors have triggered the planning and construction of power plants (PPs) of significantly increasing size: renewable energy plants include small-, large-, and very large-scale installations. Today, photovoltaic PPs and wind farms of rated power of several hundred MW and even of around 1 GW are not uncommon in several nations. Because of this trend, all the essential components of the system (e.g., PV modules, wind turbines, inverters, converters, transformers, protections, panel boards, switchgears, etc.) have been upgraded to improve the overall performance of the PPs at the lowest possible overall cost. The ultimate goal is to stay within the corporate budget limits and constraints; achieve all the Key Performance Index (KPI) targets and minimize both the worth of Capital Expenditure (CAPEX) and Operational Expenditure (OPEX). The successful achievement of these targets relies on the careful design of the layout and protection schemes of the PPs to minimize the occurrence of faults (e.g., ground faults and short circuits). Faults may occur in PPs when the basic equipment insulation fails, for example, due to overvoltages caused by lightning or switching events, insulation contamination, or other mechanical and natural causes. Faults on components may result in a severe and long-term outage of sections or even of the entire PP, which may directly impact utilities' reliability.

The design of a reliable and safe renewable energy electrical system includes the schemes of protective devices, choices about the neutral grounding, insulation coordination studies, considerations about operation and maintenance, and economic criteria. The goal of the present chapter is to address some relevant aspects and briefly review some significant issues in the above matter.

5.2 Power Collection Systems

Renewable PPs require a collection system that gathers the power produced by wind turbines and/or PV central conversion stations and brings it to a central collection point (CCP), which then ties into the power grid through the transmission system (TS) [1] (Figures 5.1 and 5.2). Usually, the collection system is designed to function in a.c. at medium voltage (MV), however, d.c. technology may also be employed [2, 3], especially for large offshore wind farms (OWFs) [4, 5], which are installed at large distances from the coast (i.e., distances equal to or greater than 100 km), where reactive power issues may make the a.c. transmission through submarine cables unfeasible.

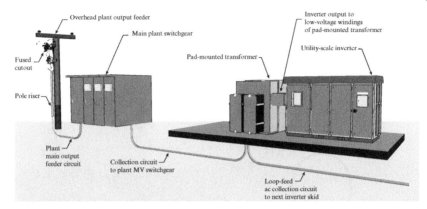

Figure 5.1 Typical MV central collection point.

The transmission system is usually made of MV cables directly buried or installed in underground PVC conduits; in some cases, it can be MV Over Head Lines (OHLs). The latter solution, if allowed by environmental laws, may be adopted in wind farm projects when the excavation of the soil is not economically convenient.

The communication system is also an integral part of the collector system and may be composed of fiber optics cables that follow the same installation modality and pathway as the MV power lines.

The collection system may be composed of several configurations: *radial layout*, *single-sided ring*, *double-sided ring*, *multi ring*, and *star layout*.

The radial configuration is the simplest, where the generating units (GUs) (i.e., wind turbines or PV systems) are connected to one feeder, forming one *string* (also called *cluster*), as shown in Figure 5.3. The power flows in one direction from the renewable energy generators to the central collection point, with the advantages of the simplicity of controls and the shorter total cable length. The radial configuration has the lowest cost but also the least overall reliability, which is its major drawback: failures of the feeder or the main switchgear will prevent all GUs from exporting power.

The ring collector configuration is used to improve the reliability of the system by employing two feeders for each cluster. The ring scheme is characterized by the presence of at least one feeder as a back up to the n in operation necessary to connect the GUs to the CCP; this configuration is referred to as $n+1$.

The *single-sided* ring is based on a radial design but features an additional feeder from the last GU to the CCP, as shown in Figure 5.4. The back-up feeder

(*a*)

(*b*)

Figure 5.2 Examples of RMU in a wind farm (a) and RMU and CCP in a PV plant (b).

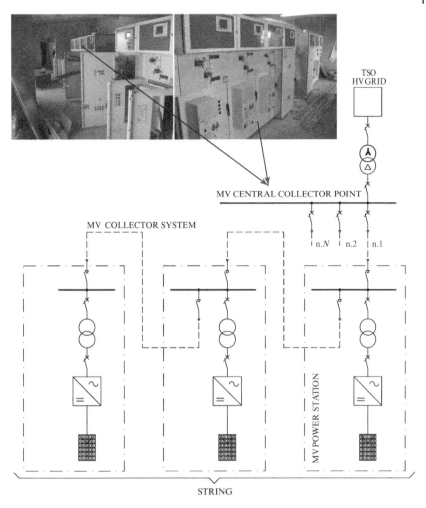

Figure 5.3 Radial collection configuration.

must be able to carry the total power generated by the string. The single-sided ring is the most expensive configuration, almost doubling the cost of the radial layout, but it is also the most reliable, being the least vulnerable to the interruption of the service. For safety reasons, maintenance personnel must be made aware of the presence of an additional source of energy and trained in the lock-out tag-out procedures.

The *double-sided* ring overcomes the cost shortcoming of the single-sided ring by using the cable of the neighbor string as the redundant feeder. With the

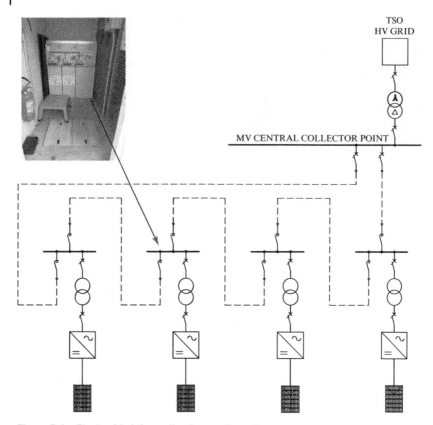

Figure 5.4 Single-sided ring collection configuration.

double-sided ring, every two radial strings are connected at their ends, as shown in Figure 5.5. If a fault occurs, any string must be rated to deliver the power of two strings. This configuration is estimated to be around 60% more expensive than the radial design.

The multi-ring shown in Figure 5.6 is conceived to divide the power generated in a string, in the case of failure of a feeder, among the other strings, so that there is no need to upgrade each string capacity, as in the case of the double-sided ring. This configuration has around 25% lower losses than a typical radial array; it is more reliable, and its cost is estimated to be around 20% higher than the radial design

In the *star* configuration (Figure 5.7), each single PV/wind generator is connected directly to the main collector. Commonly, this collector is in the center of the plant so that the lengths of the cables are optimized, and they

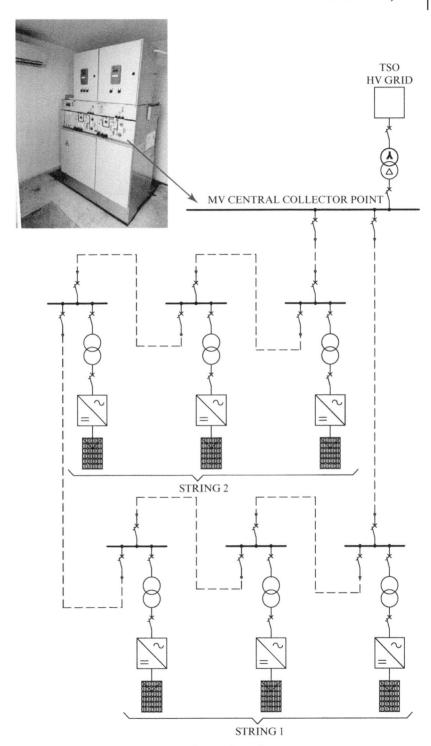

Figure 5.5 Double-sided ring collection configuration.

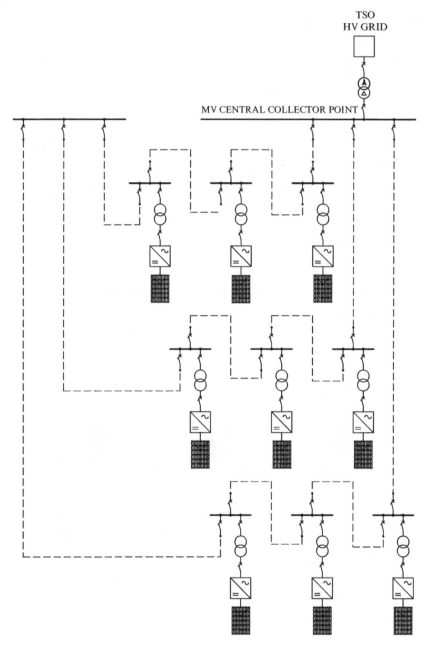

Figure 5.6 Multi ring collection configuration.

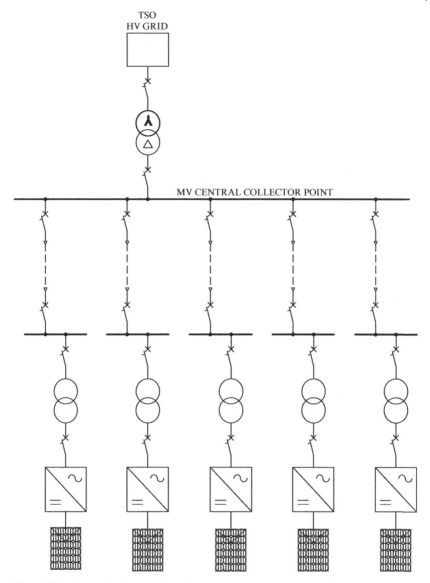

Figure 5.7 Star collection configuration.

have the same losses. This solution offers a higher reliability than the other configurations; however, the downside is the need of one feeder for each generator, which greatly increases the total cost of the switchgear at the central collection point.

When using ring configurations, the major design choice is on how to operate the ring: *normally open* or *normally closed* [6]. The normally open ring configuration usually features one (or two) network switch(es), which, from an electrical point of view, split the ring into two radial feeders. The presence of the open point simplifies the protection scheme: all the ring main units (RMUs) (Figure 5.2) are equipped with two network switches, and only the switchgear at the CCP is equipped with line feeder circuit breakers with overcurrent protection relays (e.g., 50, 51, 50N and 51N) that protect the two radial strings, without introducing any selectivity issue. When the feeders are connected to an HV/MV substation, since the point of coupling occurs with the high-voltage *transmission system operator* (TSO), and not with the MV *distribution system operator* (DNO), more advanced protective relays are required, including: 27 (undervoltage), 32 (directional power), 50BF (Breaker failure), 59 (overvoltage), 67N (directional earth fault), 74TCM (trip coil monitor), 81 (frequency – 81H overfrequency and 81L underfrequency). The drawback is that any MV fault causes the tripping of the primary feeder breaker, resulting in the disconnection of a large number of GUs, which reduces the reliability of the system. Furthermore, the fault detection may be time consuming, which may increase the outage of entire sectors of the plant.

The operation of normally open rings is not particularly complex; however, attention should be paid to avoid the accidental closing of the ring. From a more practical point of view, since the MV switches feature a three-position design (i.e., close, open, and ground), a proper captive-key interlock must be implemented. Captive-key interlocks employ locks and keys for the sequential control of switching devices to enable safe operations. Keys that are either captive or released in predetermined order are transferred among the interlocked switches.

Functional interlocks are incorporated in MV functional units and are dedicated to the operation of the apparatus located in the units only. These interlocks generally consist of specific mechanical devices, but they may also require the operation of keys.

Interlocks between different MV functional units may be implemented by means of keys transferred from one equipment to another, when they are released (Figure 5.8). Captive-key interlocking schemes are widely used as a measure of safety against the incorrect sequence of operations by personnel. The sequence should allow to operate the normally open switch without closing the ring or accidentally ground the main energized circuit.

To increase the personnel safety, two open points may be used across a line between adjacent GUs: in such case, the operating procedure should require to periodically change the position of the open line to avoid the insulation deterioration of the deenergized MV cable due to possible moisture accumulation.

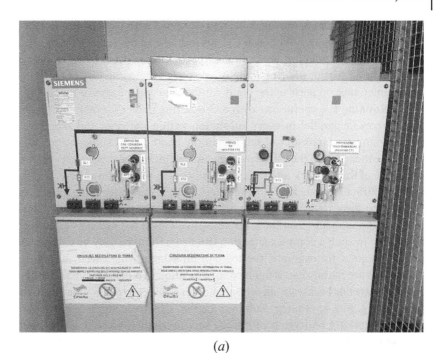

(a)

(b)

Figure 5.8 RMU (a) and CCP (b) switchgear with trapped-key interlocks.

The main advantages of ring collectors equipped with protections and circuit-breakers at the in-out of each GU substation are: the service continuity, supported by the isolation of the faulted part of the network only, allowing the remaining part of the ring to be in operation; the possibility of carrying out maintenance on parts of the plant without causing outages or plant stoppages. Running the operations with a closed ring makes it possible to have two sources in parallel in every substation within the ring, and therefore, virtually, to never have out-of-services in the plant due to faults in the ring. On the other hand, the disadvantage is the increased costs due to the extension of the network and the complexity of the protection system, which makes it difficult to quickly find and safely exclude the faulty zone. In fact, as opposed to the case of standard radial networks, the power can flow from different directions to feed the fault, thus involving more protective devices. Selectivity in such a complex situation can be guaranteed only by using special techniques, for example, the logic of *directional zone selectivity*. These techniques are mainly based upon communication of blocking signals, either in the forward or backward direction, to other devices in the system through hard wiring, bus or ethernet communication. Circuit breakers sensing the fault will start disconnecting within the set selectivity time, and at the same time send out blocking signals to the other of the circuit breakers in the network. If the circuit breaker fails to open, the blocked circuit breakers will operate after a given time delay, in accordance with the blocking signal. Thanks to the protective device communication, selectivity becomes possible in ring and meshed networks, as well as in systems with more than one power source.

In utility scale PV systems, where no particular requirements and constraints dictated by the site topography may be present, the GUs can be subdivided among cluster of equal nominal power, the ring main units feature two main entrances and the feeder of the MV/LV transformer be connected to the GUs (Figure 5.2).

Conversely, in wind farms, the clusters are usually *irregular* since the GUs are scattered over a wide area where the local topography has tight constraints (Figure 5.2). In this case, each cluster has its own nominal power, and some RMUs can be equipped with more feeders for future radial connections. Examples of regular PV and irregular wind farm clusters are reported in Figure 5.9. As shown in Figure 5.9a, in PV utility scale plants, three winding transformers are often used: the transformer has two windings for low voltage (LV) to connect two inverters, and the third winding for the MV connection to the RMU. Figure 5.2 shows some examples of RMUs and CCP in wind farms and PV systems.

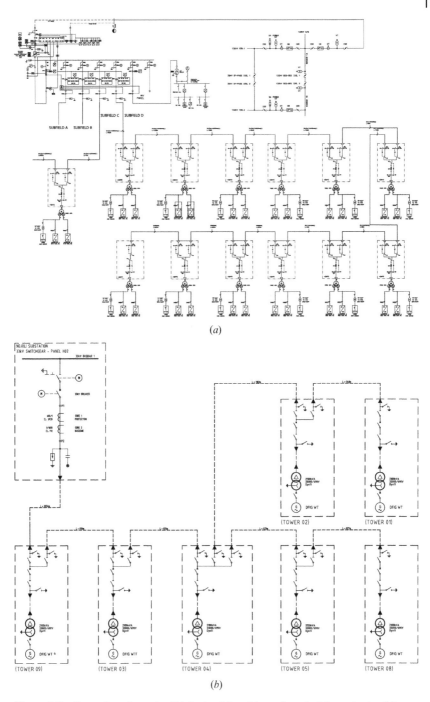

(a)

(b)

Figure 5.9 Examples of regular PV cluster (a) and irregular wind farm cluster (b).

5.3 Cable Connections

The connection between GUs is generally performed with single-core MV cables directly buried (in flat formation or trefoil formation) or in underground plastic conduits. For safety reasons, the presence of directly buried cables must be properly indicated with warning signs (see Figure 5.10). Cables with advanced protection features may be used to offer better mechanical protection than traditional metal-armored cables, while still keeping the functional advantages of unarmored cables. Available on the market are cables equipped with a polymeric protection that can absorb the kinetic energy of a shock by deforming on impact. No residual energy is therefore left to damage the sensitive parts of the cable, such as insulation and screens. Metal armoring is not equally efficient, as part of the energy of a mechanical shock is transmitted to the inner layers of the cable, potentially compromising the insulation's integrity.

In large power plants connected to the power grid, particular attention must be paid to the connection between the plant and the end-user substation. This connection may be several kilometers long and can be realized in MV or HV (in some cases via overhead power lines), depending primarily on the length and the rated power (see Figure 5.11).

The feasibility of a new renewable energy plant is determined by the capability of the power network to accept the new energy production. Criteria for feasibility include the identification of an existing electric substation, or the construction of a new one, to which the plant can be connected and the identification of a suitable right-of-way (i.e., a specific route through properties belonging to others) through which to route the new connection line. The design of the new connection infrastructure between start and end points is the real critical issue in the authorization process, rather than the design of the power plant itself.

Local populations may be hostile to the routing of new power lines because they do not perceive any direct benefit from their installation and have concerns about their health, landscape, and property values. Overhead power lines are large linear elements in the landscape and their scale is usually much larger than that of objects in their proximity (e.g., houses and trees). Their major effect is, therefore, likely to be the visual intrusion of the towers in the corridors through which the lines are routed. In addition, during the construction of the plant, and their connection, sites of natural interest may be disturbed due to the required new access tracks and their associated maintenance (Figure 5.12).

(*a*)

(*b*)

Figure 5.10 Examples of connection with directly buried cables.

(a)

150 kV overhead power line to the transmission grid

HV/MV power transformer of the wind farm

(b)

Figure 5.11 Connection infrastructure of a wind farm: MV connection to the HV/MV end-user substation (a) and overhead power line to the HV transmission grid (b) from the end-user substation.

Building new plants and transmission lines is possible if the negative impacts on people and the environment are effectively minimized, while ensuring safety and reliability. Suitable energy models, ecological assessment models, and multi-criteria approaches have been proposed with great potential for interlinking. A comprehensive framework for the assessment of environmental impacts is provided by the Strategic Environmental Assessment (SEA), which is developed for environmental impact analysis for policies, plans, and programs [7].

When realizing the connections of renewable energy plants to the power grid, obstacles such as small rivers, roads, and aqueducts may be present.

Figure 5.12 Access tracks used for the construction of a wind farm.

Figure 5.12 (*Continued*)

In some cases, cable trays or ladders with brackets secured to existing bridges may be used (Figure 5.13). Alternatively, *horizontal directional drilling* (HDD) may be often used [8 11]. Trenchless engineering is an increasingly common method of construction of pipelines, power lines, and other linear underground structures. The HDD process begins with a small pilot hole under the crossing obstacle with a continuous string of steel drill rods. When the bore head and rod emerge on the opposite side of the crossing, a special cutter (i.e., the back reamer) is attached and pulled back through the pilot hole. The purpose of the reamer is to enlarge the hole and achieve the desired hole diameter for the conduit.

Before performing HDD, a comprehensive geotechnical study is usually conducted to identify soil formations at the potential bore site. The purpose of the investigation is not only to determine if directional drilling is feasible but also to establish the most efficient way to accomplish it. Furthermore, the site must be previously surveyed by the ERT (Electro Resistivity Tomography) or Ground Penetrating Radar (GPR) imaging to identify landfill boundaries and other buried objects.

Municipalities may have particular concerns related to magnetic fields emitted by buried cables. For this reason, cables may be screened with shields of high permeable material, which increases the installation costs. In lieu of such screens, compensated twisted MV cable structure (referred to as twisted

Figure 5.13 Cable tray with brackets secured to an existing bridge.

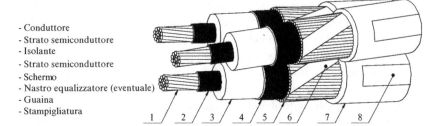

- Conduttore
- Strato semiconduttore
- Isolante
- Strato semiconduttore
- Schermo
- Nastro equalizzatore (eventuale)
- Guaina
- Stampigliatura

Figure 5.14 Twisted three-core cable, or triplex.

three-core cable, or triplex) may be used (see Figure 5.14). Analytical methodologies are available in the literature for the calculation of the magnetic field generated by a twisted configuration of conductors [12, 13].

5.4 Offshore Wind Farm

Electrical layouts for offshore wind farm (OWF) are commonly classified into six categories: small a.c., large a.c., a.c./d.c., small d.c., large d.c., and series d.c. wind farms [1, 14–16].

In small a.c. wind farms (Figure 5.15a), the medium voltage alternating current (MVAC) collector system is used for both connecting all turbines in a cluster together and transmitting the generated power to the CCP, which is usually placed onshore. The main advantage of this layout is that it does not require an offshore substation.

In the large a.c. wind farm (Figure 5.15b), turbines are clustered within the MVAC collector system. Then, the collector system is connected to a high voltage alternating current (HVAC) transmission system through an offshore substation, in which the transformer, switchgear, and auxiliary services are installed.

The layout of an a.c./d.c. wind farm (Figure 5.15c) is very similar to the previous a.c. wind farm, except for the transmission system. Wind turbines are arranged in clusters, and the wind generators produce a variable frequency and variable magnitude a.c. voltage, which needs to be processed based on the power grid frequency and voltage. A d.c. link converter, comprising a generator-side rectifier and a grid-side inverter, is normally used to condition the wind generator's outputs. The voltage is stepped up to the level required by the medium voltage collection bus, which is then further stepped up to the transmission level and fed to a HVDC converter for the d.c. transmission. At the onshore substation, the d.c. power is converted back to a.c. for the power grid connection; this arrangement requires a large number of conversion stages. In this case, a HVDC link connects the OWF to the onshore grid, therefore, submarine high-voltage cables and the CCP will feature HVDC technology.

Local wind turbine grid and transmission system

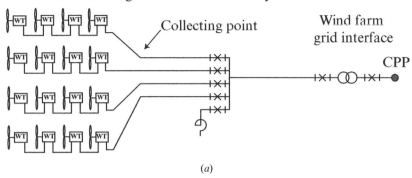

(*a*)

Local wind turbine grid and transmission system

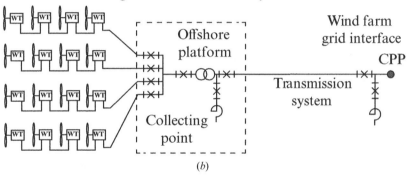

(*b*)

Local wind turbine grid and transmission system

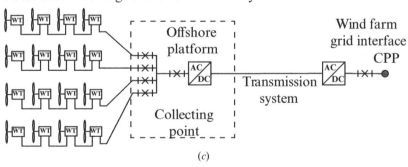

(*c*)

Figure 5.15 General layouts for offshore wind farms: (a) Small a.c. wind farm; (b) large a.c. wind farm; (c) a.c./d.c. wind farm; (d) large d.c. wind farm with two transformation steps.

Local wind turbine grid and transmission system

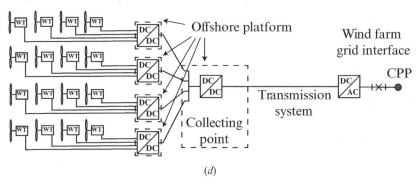

(*d*)

Figure 5.15 (*Continued*)

In the small d.c. wind farm, each wind energy conversion system (WECS) requires a rectifier to provide its output in d.c. The transformer in the wind farm grid interface is replaced with a d.c.-d.c. transformer and an inverter. This d.c. transformation step could be unnecessary depending on the medium voltage d.c. (MVDC) output of the turbines. In this case, the power is collected first in d.c. and then a d.c.-d.c. converter steps up the voltage to the transmission level (Figure 5.16a). Since the number of converters is lower and the transformers are eliminated, this configuration is more efficient. However, only two voltage levels are used and, consequently, the distribution level strongly depends on the generator voltage. The advantage of this power collection is the lowest losses because of the short distances within the wind farm. However, the required voltage transformation may not be achievable with one single stage of d.c.-d.c. conversion.

The large d.c. wind farm layout (Figure 5.15d) is conceptually similar to that of the large a.c. wind farm. The major difference is that one or two voltage transformation steps may be required among the offshore wind turbines and the transmission system (i.e., two step-ups).

In the two step-ups method (Figure 5.16b), two d.c.-d.c. converters are employed: the first steps up the voltage after each turbine up to the medium-voltage level; the second, steps up to the transmission level after the power is collected. The use of high-power d.c.-d.c. converters may lead to a significant reduction in the overall system size and weight, as well as to lesser construction and installation costs for the wind turbines and substation platform. Advantages of this configuration are the direct step-up of the voltage downstream the turbine, which leads to reduced cable losses at the distribution level, and the individual control of the voltage. One drawback of this configuration is the additional d.c.-d.c. converter, which add extra costs, losses, and a lower efficiency.

In the turbine step-up method (Figure 5.16c), single d.c.-d.c. converters are connected directly to each turbine. Hence, only two levels are used and the

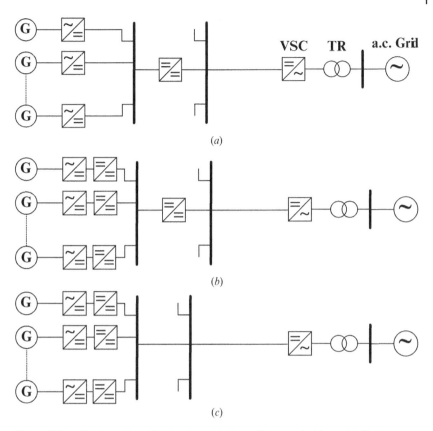

Figure 5.16 Configurations for the d.c. grid of an offshore wind farm: (a) Cluster step-up; (b) two step-ups; (c) turbine step-up.

losses at the distribution level are reduced due to the high voltage. Nevertheless, the output rectifier of the d.c.-d.c. converter must be designed for the transmission voltage with a quite low-power level, which could reduce the efficiency of these converters.

In the series d.c. wind farm, the WECSs are connected in series in order to obtain a voltage suitable for the transmission without an offshore substation.

In the past, a.c. transmission lines with high voltages in the range of 145–245 kV were the preferred technical solution for the majority of the OWFs, generally to cover not farther than a few tens of kilometers from the shore. In recent years, high-voltage d.c. transmission with voltage source converters (VSCs) has been the preferred method for the connection of long-distance OWFs [17], based on the idea that break-even distance for such power links is about 80 km [18, 19]. As explained by Lauria *et al.* [20–22] such break-even distance had been determined based on the hydraulic constraints of self-contained fluid-filled HV a.c.

cables, which were the state-of-the-art cables before the introduction of HV XLPE insulation in the 1990s, as well as by the intrinsic limit to the length of a.c. cable due to reactive power considerations.

The aforementioned limits do not affect HV d.c. lines, since most HV d.c. cables do not contain liquid insulation. In addition, VSCs HV d.c. have significant advantages: active and reactive power controls, frequency decoupling between OWF and onshore grid, and black-start capability [23]. However, the technology of HV d.c. voltage source converters is not yet fully mature: the service experience is limited, and offshore converter stations have notably experienced problems. Meanwhile, HV a.c. cable technology underwent significant improvements with the adoption of XLPE insulation, which eliminates length constraints related to oil circuits. The derating effect of reactive power is mitigated by the smaller dielectric constant and the greater thickness of the XLPE insulation, compared with paper-oil insulation, yielding a smaller capacitance; this is in addition to higher operating temperatures of the XLPE, which can allow a higher ampacity.

In the field of submarine transmission, applications are mostly at voltages not exceeding 245 kV; 420 kV – 50 Hz XLPE submarine cables are, however, currently in service. Thanks to technological advances, the maximum feasible length of a.c. cable lines has substantially increased. To fully exploit such increased length, the lines must be operated with a symmetrical reactive power profile, through appropriate terminal voltage control, which maximizes the active power transfer capability and minimizes Joule losses. Further extensions of a.c. transmission distances may be obtained with the installation of reactive compensation at intermediate locations along the route: appropriate compensation allows to actually multiply the maximum line length. Intermediate compensations of a HV and extra HV submarine cables require additional offshore platforms, which might be feasible in the relatively shallow seabed of the North Sea. Manufacturing and installation costs of a.c. submarine cables have been reduced thanks to the introduction of three-core XLPE-insulated cables, which employ galvanized steel armors, inexpensive if compared to the copper armor, necessary to keep power losses of single-core a.c. cables to an acceptable level.

5.5 Distributed Energy Resources: Battery Energy Storage Systems and Electric Vehicles

The integration of intermittent and stochastic renewable energy sources (RES) into the existing power grid may challenge the safety of operators and the stability of the energy supply and the non-dispatchable generation [24]. Battery energy storage systems (BESS) and electric vehicles (EV) play an increasingly significant role in allowing the penetration of RES, managing outages, optimizing dispatchability, reducing uncertainty, and helping reduce peak loads [25]. Generally, *distributed energy resources* (DERs) are customer-sited technologies that generate, or control, power. DERs refer to electric power generation resources that are directly connected to MV/LV/HV distribution systems, rather than to the bulk power transmission systems [26]. DERs include both generation units (which is under the umbrella term of *distributed generation*), such as wind turbines, photovoltaics, and also energy storage technologies like batteries, flywheels, superconducting magnetic energy storage, and electric vehicles, to mention a few.

Battery storage equipment may basically be of three types. *Battery modules* are simply one or more cells linked together, and they may also be equipped with electronics for monitoring, charge management, and/or protection. *Pre-assembled battery systems* comprise one or more cells, modules or battery system, and auxiliary supporting equipment, such as a battery management system and protective devices, and any other required components as determined by the equipment manufacturer. Last, but not least, the *pre-assembled integrated battery energy storage systems* (BESS) are manufactured as a complete integrated package of one or more cells, modules or battery system, protection devices, power conversion equipment (PCE), and any other required components as determined by the equipment manufacturer.

Various battery technologies are currently available for BESSs, and some have been in use for many years, while others have been developed only recently. Some of the common battery technologies available on the market are lead-acid, nickel cadmium, lithium ion, nickel metal hydride, sodium ion, sodium sulfur, and vanadium redox flow.

The real-time management of BESSs requires the deployment of two control centers: (i) a battery management system (BMS) [27], which is the brain behind the battery pack, manages its output, the charge and discharge and provides notifications on its status; (ii) an EMS [28], which interacts with grid-level applications to manage the power flow of DERs with an effective strategy to balance economic and technical issues. The continuous progress in battery technology is accelerating the transition of BMS/EMS from mere monitoring/control units to a multifunction integrated unit, the battery energy management system (BEMS).

State-of-the-art BMSs mainly react to symptoms and have a rudimentary knowledge of the battery state, and they operate the BESS rather inefficiently. For the implementation of more effective controls, the BMS needs to measure field variables, for example, cell voltages, current, and temperature. The choice of these physical quantities, and how to measure them, is of paramount importance for the determination of state of the BESS, which is represented by state functions [29], for example, state of charge, of health, etc. Methodologies to assess state functions fall into four categories [30]: direct measurements, book-keeping, adaptive and hybrid methods. By knowing the state variables, the BMS can provide safety-related functions to protect the BESS and optimize its utilization. This is accomplished by mean of a thermal-electrochemical equivalent circuit model (ECM), usually obtained through model-order reduction (MOR) techniques. When the application of the BESS is known, necessary requirements on the sensors can be derived as well. A very active research presently focuses on self-learning models based on artificial intelligence, which allow the BMS to predict the battery behavior [31, 32].

The role of the EMS is to guide the energy flow through the supply system, and a number of strategies have been proposed on this hot topic [33, 34]. The EMS is required to find a balance among several competing technical/economic/environmental objectives: enhance system resilience, prioritizing self-consumption accounting for prices, maximizing RES penetration, and enhancing integration in hybrid systems. Cooperative strategies include energy shifting, peak shaving, load following, etc. EMS requires artificial intelligence to perform advanced functions as forecasting future information, manage distributed BESS in a cooperative way, handle uncertainties: nonlinear programming, heuristic algorithms, fuzzy logic, artificial (recursive) neural networks, and their combination have been eminent choices. When distributed multi-agent architectures have been used, issues arose, such as time lags from sensors, synchronization of updating systems, and deficit between real time and forecasting [35].

Most battery cells feature low voltages and, therefore, the risk of electric shock is limited; however, some large battery banks may produce more than 120 V d.c. Batteries could be a serious safety risk for users and technicians, if incorrectly installed and operated, potentially leading to electric shock, fire, flash burns, explosion, or exposure to hazardous chemicals and released gases.

Personnel should be protected from the dangers of electric shock by ensuring that live conductors are effectively insulated or protected; access to areas where hazardous voltages are present is controlled; appropriate warning signage is displayed. The battery bank must be electrically isolated during any work on or near it. The isolation of batteries requires the use of secure insulating caps or barriers of the terminals.

Battery installations should be designed to eliminate or reduce the risk of fault currents associated with battery terminals or short circuits to the battery stands or trays. Battery stands or trays should be insulated, and access to battery terminals, inspection caps, or charge indicators should be sufficient to allow effective and safe maintenance. Large battery systems should be installed in a cool, well-ventilated area away from ignition sources, better if in a dedicated battery room with minimal presence of other equipment and services. A battery has sufficient energy to cause an arc flash, which produces temperatures above 12,000°C, and can melt metal or causing fires and explosions. Higher battery energy storage capacities have a higher risk of severe arc flash. Arcing faults may cause the catastrophic failure of battery cell enclosures unless the fault currents are quickly removed by properly rated protective devices. The short circuit ratings of BESS are specified by the manufacturer. It is imperative that overcurrent protection devices (i.e., fuses or circuit breakers) be correctly selected and installed to withstand such currents.

Most lead-acid batteries generate hydrogen and oxygen when charging. Other battery types may release flammable gases and need adequate ventilation to prevent explosive atmospheres, fire, or risk to users. Lithium-ion batteries do not produce exhaust gases during normal operation, but they can produce flammable and toxic gases during faults. Fire and explosions can result from component failure, short circuits, loose connections, or human error during installation or maintenance work.

Battery casings can degrade or be damaged by impacts, and they can also rupture as a result of excessive temperatures and pressure generated by the chemical reaction from over-charging, or following a short circuit. Electrolyte (fluid or gel) can leak from a ruptured casing, resulting in toxic fumes, burns, corrosion, or explosion.

A Type B RCD should be installed on the a.c. side of the PCE to be sensitive to fault currents with a low-level ripple.

Considered as a pooled resource, the growing number of electric vehicle batteries could provide a wide range of valuable grid services, from demand-response and voltage regulation to distribution-level services, without compromising driving experience or capability [36]. Electric utility companies, featuring an increasing amount of renewable energy sources in their portfolio, can use new communications and control technologies, together with innovative tariffs and incentive structures, to tap the sizeable value potential of smart electric-vehicle charging to benefit customers, shareholders, vehicle owners, and the society at large [37].

In this framework, some safety aspects should be addressed.

According to the IEC 61851-1 "*Electric vehicle conductive charging system – Part 1: General requirements,*" electric vehicles (EVs) can be charged in four different modes (see Figure 5.17).

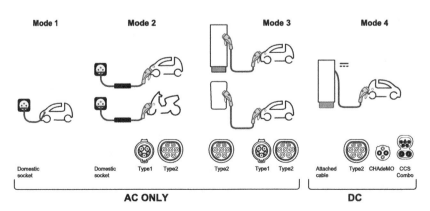

Figure 5.17 Charging modes.

Mode 1 consists of the direct connection of the EV to the regular dwelling unit current socket through a cable and plug (the rated values for current and voltage shall not exceed 16 A and 250 V, 3.7 kW, in single-phase, while 16 A and 480 V, 11 kW, in three-phase); at 230 V, it may normally take 6–8 hours to fully charge an EV battery. Mode 1 does not require any supplementary pilot or auxiliary safety system, referred to as *electric vehicle supply equipment* (EVSE), or any control pins. As a protection against electric shock, Mode 1 requires a residual current device (RCD), which requires a protective grounding conductor. The absence of a grounding system in dwelling units is one of the reasons why Mode 1 is not allowed in some countries.

Mode 2 requires the presence of an inline *control box* in the charging cable for control pilot function. The control box necessitates a specific pilot wire and a pulse width modulation (PWM) communication channel with the vehicle. The charging station communicates to the EV the grid availability by means of a frequency-modulated signal, and the EV adapts the load by returning its own status through a voltage value. For vehicles without PWM, the circuit works in simplified mode, and the station limits the charging current to 16 A. Mode 2 can be used with both residential and industrial receptacles (the rated values for current and voltage shall not exceed 32 A and 230 V, 7.4 kW in single-phase while 32 A and 400 V in three-phase, 22 kW). At 230 V, it usually takes 6–8 hours to fully charge the EV.

Mode 2 connectors require a control pin only on the vehicle, according to IEC 61851-1. On the supply network, a control pin is not required, and the control function is powered by the control box. Between pilot and ground, a 1 kΩ impedance flags spikes running through the pilot-ground loop and acts as circuit breaker. Mode 2 is allowed only for private use and subject to severe restrictions in public areas.

Mode 3 is the most widespread charging system: it requires the EV to be charged through the EVSE permanently connected to the electrical network, with control pilot and safety functions. This is the mode of wall-boxes, commercial charging points, and all automatic charging systems in a.c. The charging cable runs from the EVSE to the vehicle, allowing PWM communication. The control pilot function confirms correct connection, integrity of ground, and current capacity of the charger. Depending on the capacity, a resistor is fitted between the control pilot contact and the ground, the value of which identifies the cable size. In this case, the charging can take place at different power levels; therefore, it can be either slow or fast. A 400 V and 63 A three-phase source allows a full charging of the battery in under one hour. It is allowed in public areas; at home and in private areas, it requires a wall-box.

Mode 4 is the only charging mode directly in d.c. It requires an off-board converter which provides d.c. supply to the EV (a.c./d.c. conversion) and which has a control pilot and safety function. The supply voltage is 400 V and the top current is 200 A.

The Japanese Company CHAdeMO invested more than any other company towards the development of this charging mode worldwide. Its high voltage and current require great attention to safety and the correct use of appropriate connectors and cables. A range of control and signal pins are required to obtain fast charging comparable to that of Mode 3.

As to the connectors, a.c. charging systems accept four different types of connectors. Type 1 (see Figure 5.18a), also called standard SAE J1772 (or Yazaki), is mostly most used in the American and Asian markets, and it has been

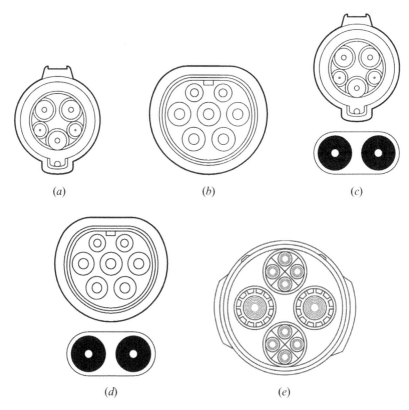

Figure 5.18 Plugs: (a) Type 01, (b) Type 2, (c) Type CCS-1, (d) Type CCS-2, (e) CHAdeMO.

designed under Japanese standards. This connector, placed only on the EV (i.e., at the inlet), has five terminals and can support two different levels of charge in the alternating current: one of 16 A for a normal or slow charge and another that reaches up to 80 A for a fast charge.

Type 2 (see Figure 5.18b), also known as Mennekes, is mostly adopted in Europe, and it is designed according to Std. IEC 62196-2. This connector has seven terminals and offers two levels of charge: through three-phase direct current at 63 A and 44 kW for Mode 3 or semi-fast charge and through a.c. on single phase at 16 amps and 3.7 kW for Mode 2 or slow charge. It is placed both on the EV and on the charging station.

Type 3A (or Scame) is the preferred connector for small vehicles with less than 3 kW power, such as scooters and motorcycles. It requires the fourth pin to monitor the pilot circuit and verify continuity of the protective conductors.

Type 3 C, originally intended for electric vehicles with ratings above 3 kW, is no longer used.

Mode 4 charging allows two different sockets: the CCS or Combo and the CHAdeMO.

The CCS (*Combined Charging System*), also called *combo*, is a connector that is trying to become the standard solution that allows both charging modes, slow and fast. It combines both a.c. charging over longer downtimes and fast d.c. charging in a single connection. The connector is composed by the marriage of two connectors, one that offers a.c. through Yazaki (in CCS type 1 – see Figure 5.18c) or Mennekes (in CCS type 2 – see Figure 5.18d) and another that offers d.c. through two contacts or terminals. Therefore, the connector allows to charge the EV in modes 2, 3, and 4. With five terminals, the maximum power at which the CCS can operate in a.c. is 44 kW and up to 350 kW in d.c.

The CHAdeMO (see Figure 5.18e) connector has been designed by the Japanese Industry to perform Mode 4 fast charging using d.c. and can withstand powers ranging between 50 kW and 400 kW. The high charging voltage and power require safety features, and therefore it is equipped with 10 pins, which makes it the largest diameter connector currently in the market.

Conversion systems are normally present in EVSE, such as control box and off-board converters, which present new challenges to the electric safety of users. EV charging equipment increase the risk of electric shock under several aspects: risk of discontinuity of the protective conductor in plugs; risk of mechanical damage to cable insulation, for instance, due to crushing by rolling of vehicle tires, repeated operation, etc.; risk of access to live parts of the Class 1 charger in the EV as the result of deterioration or even destruction of the basic protection, due to accidents, car maintenance, etc.; wet or saltwater wet environments, for example, rain or snow on the electric vehicle inlet, etc. Specific requirements for safety and design are provided in Std. IEC 60364

"Low-voltage electrical installations – Part 7–722: Requirements for special installations or locations – Supplies for electric vehicles."

The standard dictates a dedicated circuit to transfer the energy from/to the EV. In TN grounding systems, a circuit supplying a connecting point must not include a PEN conductor. Additionally, it should be verified whether electric vehicles using the charging stations have limitations related to specific grounding systems (e.g., some vehicles do not support IT grounding systems).

According to the IEC 60364, additional protection with an RCD with a rated current of 30 mA is required, which must disconnect all live conductors. The RCD must be installed upstream from the EV charger (at the main supply or distribution panel), even if the EV charger has its own RCD built in. This is because the external RCD will protect a person in the event of a broken cable between the distribution panel and the EV charger (e.g., a person accidentally cuts the supply cable with a power tool).

Type A RCDs have been initially used. However, the installation of Type B RCDs has been recently encouraged by various guidelines and standards. Standard RCDs protect persons and equipment at the rated frequency of 50/60 Hz and are not specifically tested in the presence of residual currents with a relevant content of high frequency components. The new Type F RCD (where "F" stands for frequency), introduced in Std. IEC 62423, must pass specific high-frequency tests and protects people against multifrequency ground-fault currents generated by charging stations, which can cause ventricular fibrillation and electrocution. Std. IEC 60364 allows type A or type B RCDs (or better F) in conjunction with a *residual direct current detecting device* (RDC-DD) that complies with Std. IEC 62955.

The installation choice is mainly dictated by the maximum d.c. current that may flow through the EV circuit without tripping. It is 60 mA for 30 mA type B RCD (i.e., $2I_{\Delta n}$ as per IEC 62423) and 6 mA for RDC-DD (as per IEC 62955). In damp/wet environments, or due to the aging of the insulation, the leakage current is likely to increase up to 5 or 7 mA and may lead to nuisance tripping. It should be kept in mind that RCDs other than Type B are not designed to function correctly in the presence of d.c. leakage current and may be *"blinded"* if this current is too high. Their core will be pre-magnetized and saturated by the d.c. current and may become desensitized to the a.c. fault current; the RCD may no longer trip in the case of a.c. fault, leading to a potential hazardous situation.

To complete the protection, a coordinated SPD system is usually required, whose design will depend on whether the building where the EV charge station is installed is protected by an external LPS (see Figure 5.19). As general rules, for type 1 and 2 SPDs are required in the main low-voltage panel, where the dedicated circuit to the EVSE originates; an additional type 2 SPD is required in each EVSE, except if the distance from the main panel to the EVSE is less than 10 m.

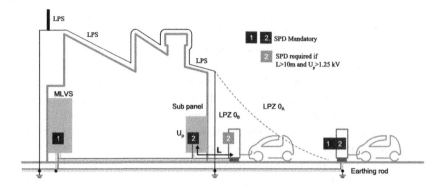

Figure 5.19 Coordinated LPS and SPD system for EV charging stations.

A type 3 SPD is recommended for the load management system, being sensitive electronic equipment. The type 3 SPD must be installed downstream a type 2 SPD, which is generally recommended in the panel where the load management system is installed.

Vehicle-to-Everything (V2X) is an umbrella term to define the use of EV batteries to provide energy services, which associate additional value to the battery asset during times of non-use. Possible operating modes are Vehicle-to-Grid, to-Building, to-Home, and to-Load [38–41] when EV batteries, respectively, interact with and provide value to the electric grid, optimize building energy consumption, optimize home energy consumption or use EV batteries as emergency back-up power, use EV batteries to provide energy to a load. EVs located *behind-the-meter* have a dual value: optimizing their own energy consumption as well as interacting with the power grid.

References

1 Madariaga, A., Martín, J.L., Zamora, I., Martínez de Alegría, I., and Ceballos, S. (2013). Technological trends in electric topologies for offshore wind power plants. *Renewable and Sustainable Energy Reviews* 24: 32–44.

2 Alhuwaishel, F.M., Allehyani, A.K., Al-Obaidi, S.A.S., and Enjeti, P.N. (2020). A medium-voltage DC-collection grid for large-scale PV power plants with interleaved modular multilevel converter. *IEEE Journal of Emerging and Selected Topics in Power Electronics* 8 (4): 3434–3443.

3 Chuangpishit, S., Tabesh, A., Moradi-Shahrbabak, Z., and Saeedifard, M. (2014). Topology design for collector systems of offshore wind farms with pure DC power systems. *IEEE Transactions on Industrial Electronics* 61 (1): 320–328.

4 Robinson, J., Jovcic, D., and Joos, G. (2010). Analysis and design of an offshore wind farm using a MV DC grid. *IEEE Transactions on Power Delivery* 25 (4): 2164–2173.

5 Jovcic, D., Taherbaneh, M., Taisne, J., and Nguefeu, S. (2015). Offshore DC grids as an interconnection of radial systems: Protection and control aspects. *IEEE Transactions on Smart Grid* 6 (2): 903–910.

6 Wadi, M., Baysal, M., and Shobole, A. (2017). Comparison between open-ring and closed-ring grids reliability. *2017 4th International Conference on Electrical and Electronic Engineering (ICEEE)*, 290–294.

7 Araneo, R., Celozzi, S., and Vergine, C. (2014). Eco-sustainable routing of power lines for the connection of renewable energy plants to the {I}talian high-voltage grid. *Int. J. Energy Environ. Eng.* 6 (1): 9–19.

8 Maria, K. (May 2020). Comprehensive risk management in horizontal directional drilling projects. *Journal of Construction Engineering & Management* 146 (5): 4020034.

9 Ledroz, A.G., Pecht, E., Cramer, D., and Mintchev, M.P. (2005). FOG-based navigation in downhole environment during horizontal drilling utilizing a complete inertial measurement unit: Directional measurement-while-drilling surveying. *IEEE Transactions on Instrumentation and Measurement* 54 (5): 1997–2006.

10 Pecht, E. and Mintchev, M.P. (2007). Modeling of observability during in-drilling alignment for horizontal directional drilling. *IEEE Transactions on Instrumentation and Measurement* 56 (5): 1946–1954.

11 Polak, M.A. and Lasheen, A. (2001). Mechanical modelling for pipes in horizontal directional drilling. *Tunnelling and Underground Space Technology* 16: 47–55.

12 Mazzanti, G., Landini, M., and Kandia, E. (2010). A simple innovative method to calculate the magnetic field generated by twisted three-phase power cables. *IEEE Transactions on Power Delivery* 25 (4): 2646–2654.

13 Ehrich, M. and Fichte, L.O. (1999). Magnetic field reduction of twisted three-phase power cables of finite length by specific phase mixing. *1999 International Symposium on Electromagnetic Compatibility (IEEE Cat. No.99EX147)*, 448–451.

14 Akay, B., Ragni, D., Ferreira, C.S., and Van Bussel, G.J.W. (2013). Investigation of the root flow in a horizontal axis. *Wind Energy* no. March 2012, 1–20.

15 Alagab, S.M., Tennakoon, S., and Gould, C. (2015). Review of wind farm power collection schemes. *Proceedings of the Universities Power Engineering Conference*, 2015-Novem.

16 Elshahed, M., Ragab, A., Gilany, M., and Sayed, M. (2021). Investigation of switching over-voltages with different wind farm topologies. *Ain Shams Engineering Journal.*

17 Xu, L., Yao, L., and Sasse, C. (2007). Grid integration of large DFIG-based wind farms using VSC transmission. *IEEE Transactions on Power Systems* 22 (3): 976–984.

18 Bresesti, P., Kling, W.L., Hendriks, R.L., and Vailati, R. (2007). HVDC connection of offshore wind farms to the transmission system. *IEEE Transactions on Energy Conversion* 22 (1): 37–43.

19 Van Eeckhout, B., Van Hertem, D., Reza, M., Srivastava, K., and Belmans, R. (July 2010). Economic comparison of VSC HVDC and HVAC as transmission system for a 300 MW offshore wind farm. *European Transactions on Electrical Power* 20 (5): 661–671.

20 Lauria, S., Schembari, M., Palone, F., and Maccioni, M. (May 2016). Very long distance connection of gigawatt-size offshore wind farms: Extra high-voltage AC versus high-voltage DC cost comparison. *IET Renewable Power Generation* 10 (5): 713–720.

21 Lauria, S., Maccioni, M., Palone, F., and Schembari, M. (2015). Cost evaluation of EHVAC offshore wind farm using intermediate shunt compensation: A parametric study. *IET Conference Proceedings* 003 (6 .)-003 (6 .)(1).

22 Lauria, S. and Palone, F. (2014). Optimal operation of long inhomogeneous AC cable lines: The Malta–Sicily interconnector. *IEEE Transactions on Power Delivery* 29 (3): 1036–1044.

23 Elliott, D. et al. (2016). A comparison of AC and HVDC options for the connection of offshore wind generation in Great Britain. *IEEE Transactions on Power Delivery* 31 (2): 798–809.

24 Carrasco, J.M. et al. (2006). Power-electronic systems for the grid integration of renewable energy sources: A survey. *IEEE Transactions on Industrial Electronics* 53 (4): 1002–1016.

25 Driesen, J. and Katiraei, F. (2008). Design for distributed energy resources. *IEEE Power Energy & Magazine* 6 (3): 30–40.

26 Akorede, M.F., Hizam, H., and Pouresmaeil, E. (2010). Distributed energy resources and benefits to the environment. *Renewable and Sustainable Energy Reviews* 14 (2): 724–734.

27 Xiong, R., Li, L., and Tian, J. (2018). Towards a smarter battery management system: A critical review on battery state of health monitoring methods. *Journal of Power Sources* 405: 18–29.

28 Shen, M. and Gao, Q. (2019). A review on battery management system from the modeling efforts to its multiapplication and integration. *International Journal of Energy Research* 43 (10): 5042–5075.

29 Wang, Y. et al. (2020). A comprehensive review of battery modeling and state estimation approaches for advanced battery management systems. *Renewable and Sustainable Energy Reviews* 131: 110015.

30 Ungurean, L., Cârstoiu, G., Micea, M.V., and Groza, V. (Feb. 2017). Battery state of health estimation: A structured review of models, methods and commercial devices. *International Journal of Energy Research* 41 (2): 151–181.

31 Wu, B., Widanage, W.D., Yang, S., and Liu, X. (2020). Battery digital twins: Perspectives on the fusion of models, data and artificial intelligence for smart battery management systems. *Energy AI*.

32 Ng, M.-F., Zhao, J., Yan, Q., Conduit, G.J., and Seh, Z.W. (2020). Predicting the state of charge and health of batteries using data-driven machine learning. *Nature Machine Intelligence*.

33 Weitzel, T. and Glock, C.H. (2018). Energy management for stationary electric energy storage systems: A systematic literature review. *European Journal of Operational Research* 264 (2): 582–606.

34 Olatomiwa, L., Mekhilef, S., Ismail, M.S., and Moghavvemi, M. (2016). Energy management strategies in hybrid renewable energy systems: A review. *Renewable and Sustainable Energy Reviews* 62: 821–835.

35 Rosato, A., Panella, M., Araneo, R., and Andreotti, A. (2019). A neural network based prediction system of distributed generation for the management of microgrids. *IEEE Transactions on Industry Applications* 55 (6): 7092–7102.

36 Howell, S., Rezgui, Y., Hippolyte, J.-L., Jayan, B., and Li, H. (2017). Towards the next generation of smart grids: Semantic and holonic multi-agent management of distributed energy resources. *Renewable and Sustainable Energy Reviews* 77: 193–214.

37 Burger, S.P. and Luke, M. (2017). Business models for distributed energy resources: A review and empirical analysis. *Energy Policy* 109: 230–248.

38 Turker, H. and Bacha, S. (2018). Optimal minimization of plug-in electric vehicle charging cost with vehicle-to-home and vehicle-to-grid concepts. *IEEE Transactions on Vehicular Technology* 67 (11): 10281–10292.

39 Liu, C., Chau, K.T., Wu, D., and Gao, S. (2013). Opportunities and challenges of vehicle-to-home, vehicle-to-vehicle, and vehicle-to-grid technologies. *Proceedings of the IEEE* 101 (11): 2409–2427.

40 Kuang, Y., Chen, Y., Hu, M., and Yang, D. (2017). Influence analysis of driver behavior and building category on economic performance of electric vehicle to grid and building integration. *Applied Energy* 207: 427–437.

41 Barone, G., Buonomano, A., Calise, F., Forzano, C., and Palombo, A. (2019). Building to vehicle to building concept toward a novel zero energy paradigm: Modelling and case studies. *Renewable and Sustainable Energy Reviews* 101: 625–648.

6

Soil Resistivity Measurements and Ground Resistance

CONTENTS

Happy is he who was able to know the causes of things.

Virgil

6.1   Soil Resistivity Measurements

The installation of renewable energy systems normally requires the geological investigation of the site to determine the values of the soil's electrical resistivity at different depths. The soil resistivity is a crucial input parameter for the correct design of grounding electrodes.

The analytical formulae so far discussed are applicable to an electrically homogeneous and isotropic half-space, with a uniform resistivity ρ. In reality,

Electrical Safety Engineering of Renewable Energy Systems. First Edition. Rodolfo Araneo and Massimo Mitolo.
© 2022 by The Institute of Electrical and Electronics Engineers, Inc. Published 2022 by John Wiley & Sons, Inc.

a grounding system design based on a single-layer soil is not generally accurate, as by nature, the soil is hardly uniform.

Table 6.1 shows the resistivity of typical soils, which may range over five orders of magnitude.

The soil body may consist of horizontal and vertical layers of different thicknesses, which may differ in texture, structure, consistency, color, and other chemical, biological, electrical, and physical characteristics. Thus, the soil may be represented by an *N*-layer horizontal model [1], where each layer may have a different resistivity. The resistivity is mainly influenced by the moisture content of the soil, the soil pH, the aggregate percentage of the different components, and the ambient temperature.

Different soil resistivity measurement techniques [2–4] have been described in literature and in technical standards (e.g., IEEE Std. 81, IEEE Std. 80), the most common technique being the *four-pin* method.

The measurement of electrical resistivity requires four electrodes (i.e., pins): two outer electrodes A and B, referred to as the *current electrodes* (Figure 6.1) inject the current *I* into the soil, and two inner electrodes M and N, referred to as the *potential electrodes*, read the resulting potential difference *V*.

The resistivity is derived through a *geometric factor K* from the measured *apparent resistance* $R = \dfrac{V}{I}$ [5].

Nowadays, all-in-one multinode resistivity and sounding/profiling equipment for geophysical studies are available on the market (Figure 6.2).

Table 6.1 Typical soil resistivity of various types of soil.

Type of soil or water	Typical resistivity (Ωm)
Sea water	2
Clay	40
Ground well and spring water	50
Clay and sand mix	100
Shale, slates, sandstone	120
Peat, loam, and mud	150
Lake and brook water	250
Sand	2,000
Morane gravel	3,000
Ridge gravel	15,000
Solid granite	25,000
Ice	100,000

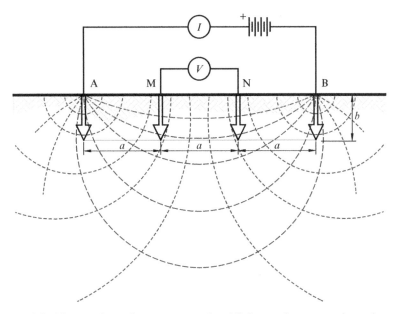

Figure 6.1 Wenner electrode arrangement. A and B denote the current electrodes, and M and N the potential electrodes.

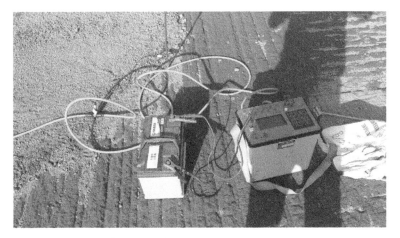

Figure 6.2 IRIS Syscal Pro equipment. (Photo courtesy of Prof. Luciana Orlando).

The measured value of soil resistivity is referred to as the *apparent resistivity* ρ_a, because it is used in the calculation of the soil multi-layer model and is not an actual value of resistivity. The apparent resistivity is defined as the resistivity of an electrically homogeneous and isotropic half-space that would yield the same ground resistance as that of a given ground electrode buried in a multi-layer soil.

If the soil is homogenous and isotropic, the apparent resistivity will be constant and correspond to the actual resistivity of the uniform soil; if the earth is non-homogenous, the resistivity is *apparent,* and when the electrode spacing is changed for each measurement, a different value of resistivity will be found.

6.2   Wenner Method

The Wenner method [6–9] is the technique typically used for soil resistivity measurements. It involves a symmetric configuration of pins (Figure 6.1): four rods placed in a straight line and equally spaced. Assuming that the depth b of the penetration of the probes into the soil is small compared to the spacing a of the four probes (i.e., $a > 10b$), so that the probes can be considered point electrodes, the potential difference V measured between the electrodes M and N is given by:

$$V = \frac{\rho_a I}{2\pi} \left[\left(\frac{1}{r_{AM}} - \frac{1}{r_{BM}} \right) - \left(\frac{1}{r_{AN}} - \frac{1}{r_{BN}} \right) \right]. \tag{6.1}$$

The terms r_{AM}, r_{BM}, r_{AN}, and r_{BN}, indicate distances between the probes.

By solving for the apparent resistivity ρ_a, we obtain:

$$\rho_a = \frac{V}{I} \underbrace{\frac{2\pi}{\left(\frac{1}{r_{AM}} - \frac{1}{r_{BM}} \right) - \left(\frac{1}{r_{AN}} - \frac{1}{r_{BN}} \right)}}_{K} = \frac{V}{I} 2\pi a = RK. \tag{6.2}$$

If the depth b is comparable to the spacing a, the apparent resistivity may be approximated by the following expression obtained through data fitting:

$$\rho_a = R \frac{4\pi a}{1 + \frac{2a}{\sqrt{a^2 + 4b^2}} - \frac{a}{\sqrt{a^2 + b^2}}}. \tag{6.3}$$

The soil resistivity is investigated through repeated measurements obtained by changing the probe spacing a and the direction x of the probe array. The repeated measurements provide a two-dimensional display of both horizontal and vertical variations in the apparent resistivity, which is referred to as *pseudo-section* (Figure 6.3).

To detect horizontal changes of soil resistivity a *profiling*, or electrical *trenching*, is performed: all four probes are moved across the surface, with the spacing a between each adjacent pair remaining the same.

To detect vertical changes of soil resistivity, a *vertical electrical sounding* (VES), or electrical *drilling*, is performed: the distance a between the electrodes

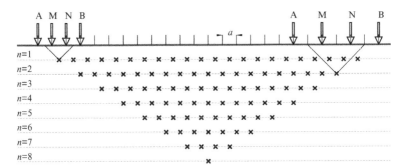

Figure 6.3 Establishment of a 2D electrical resistivity pseudo-section with the Wenner method.

is increased, while maintaining constant the location x of the center point of the array. The *continuous vertical electrical sounding* (CVES) combines both profiling and sounding to obtain a 2D resistivity data model of the soil. The standard graphic representation of Figure 6.5 shows each measured value at the intersection of two 45° lines originating at the mid-centers of the segments $\overline{AM}$ and $\overline{NB}$. When performing VES, the inter-electrode spacing a may be increased by a factor n, and an imaginary horizontal line into the soil is associated with the value of n, and a *pseudo-depth* of investigation is therefore identified.

Based on the readings of the CVES, the actual values of the soil resistivity must be assigned to each point of the soil. This requires the *data interpretation*, which is a mathematical inversion of the apparent resistivity values into the interpreted (actual) soil resistivity. This allows a mathematical representation of the soil, whose electrical resistivity may be estimated from the observed data. All inversion methods of the apparent resistivity values essentially determine a model for the soil subsurface, whose response agrees with the measured data, within certain restrictions and acceptable limits. In other words, the inversion consists of converting the volumetric soil apparent resistivity into inverted resistivity data, which represent the resistivity at an actual depth of investigation, rather than that at the pseudo-depth.

It is commonly assumed that the depth of the actual prospection corresponds to the distance between the electrodes A and M. It is important to keep in mind that the actual depth of prospection does depend on the resistivity of the layer: a conductive layer (e.g., a water pocket) may reduce the investigation depth.

The inversion process is inherently non-linear and generally requires numerical modeling. Additionally, the non-uniqueness of the solution in the inversion procedure may cause misinterpretations of the results. Methods of interpretation may include different approaches, such as approximation, direct, iterative, combined direct-iterative, and artificial intelligence.

Numerical approaches characterize the depth $h_i(x)$ and the resistivity ρ_i of each layer, to fit the measured apparent values, usually by minimizing a common object function. The number N of layers can be established a priori or can be a variable of the inversion (or optimization) process. In more sophisticated numerical approaches, the subsurface section is subdivided into cells of fixed dimensions or regions with variable depths (usually with homogeneous resistivity), and the model parameters (e.g., resistivities and boundary depths) are optimized so that the apparent resistivity values match the observed values. As an example, Figure 6.4 shows the CVES data obtained through commercial software.

If the apparent resistivity is constant for various probe spacings, the soil resistivity is fairly constant, and a uniform soil model may be used with reasonable accuracy. The IEEE Std. 80 suggests in this case to average the N measured data $\rho_{a,i}$ to obtain an equivalent uniform soil resistivity:

$$\rho \cong \frac{1}{N}\sum_{i=1}^{N}\rho_{a,i}. \tag{6.4}$$

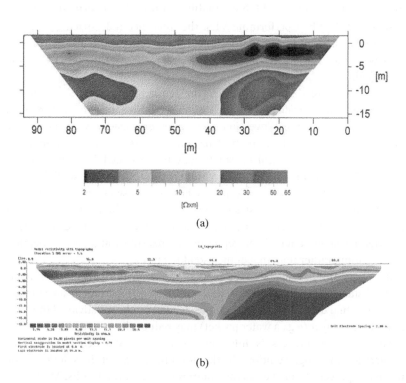

(a)

(b)

Figure 6.4 CVES data obtained through commercial software Surfer (*a*) and RES2DINV (*b*) (Photo courtesy of Prof. Luciana Orlando, Sapienza University of Rome).

Alternatively, the same standard proposes a more direct formula based on maximum and minimum apparent resistivity values:

$$\rho \cong \frac{\rho_{a,\text{min}} + \rho_{a,\text{max}}}{2}. \tag{6.5}$$

Typically, soils are hardly ever uniform, and the apparent resistivity varies with the spacing of the probes. Large variations in the readings of the apparent resistivity (i.e., greater than 30%) indicate that the soil is non-uniform, and a N-layer soil model should be used.

Herein, we discuss the two-layer soil model, which is widely used and proved to be effective in the majority of the grounding system designs. A three-layer stratified soil model may be employed when the results with the two-layer soil are deemed inadequate.

The two-layer soil model has an upper layer of a finite depth H and resistivity ρ_1, and a lower layer of infinite depth and resistivity ρ_2 (Figure 6.5).

The ground potential at point P generated by a point source S can be obtained by applying the *image principle* [10], which superimposes the potential generated by the actual source and the potentials generated by the images of the source. If the soil is multi-layered, the number of the images is infinite.

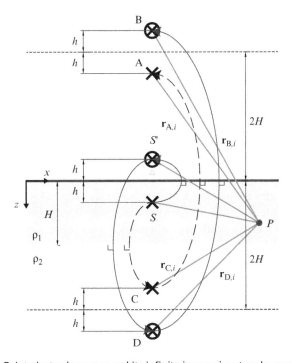

Figure 6.5 Point electrode source and its infinite images in a two-layered soil model.

The ground potential $V(P)$ is:

$$V(P) = \frac{I\rho_1}{4\pi}\left[\frac{1}{r} + \frac{1}{r'} + \sum_{i=1}^{\infty} K^n\left(\frac{1}{r_{A,i}} + \frac{1}{r_{B,i}} + \frac{1}{r_{C,i}} + \frac{1}{r_{D,i}}\right)\right]. \tag{6.6}$$

K is the reflection coeffcient at the boundary of the two layers, defined as $K = \dfrac{\rho_2 - \rho_1}{\rho_2 + \rho_1}$.

Depending on where the point source S and the point P are located, the potential can be calculated as:

$$V(P) = \begin{cases} \left|\dfrac{I\rho_1}{4\pi}\left[\dfrac{1}{\sqrt{\xi^2 + (h-z)^2}} + \dfrac{1}{\sqrt{\xi^2 + (h+z)^2}} + \right.\right. \\ \quad + \sum\limits_{i=1}^{\infty} K^i\left(\dfrac{1}{\sqrt{\xi^2 + (2iH + h - z)^2}} + \dfrac{1}{\sqrt{\xi^2 + (2iH + h + z)^2}} + \right. & z(S)\wedge z(P) < H \\ \quad \left.\left.\left. + \dfrac{1}{\sqrt{\xi^2 + (2iH - h - z)^2}} + \dfrac{1}{\sqrt{\xi^2 + (2iH - h + z)^2}}\right)\right]\right| \\[2em] \dfrac{I\rho_2}{4\pi}\left[\dfrac{1}{\sqrt{\xi^2 + (h-z)^2}} - \dfrac{K}{\sqrt{\xi^2 + (-2H + h + z)^2}} + \right. & z(S)\wedge z(P) > H \\ \quad \left. + (1 - K^2)\sum\limits_{i=0}^{\infty} K^i \dfrac{1}{\sqrt{\xi^2 + (2iH + h + z)^2}}\right] \\[2em] \dfrac{I\rho_1}{4\pi}(1 + K)\sum\limits_{i=0}^{\infty} K^i\left(\dfrac{1}{\sqrt{\xi^2 + (2iH + h + z)^2}} + \dfrac{1}{\sqrt{\xi^2 + (2iH - h + z)^2}}\right) & z(S) < H,\ z(P) > H \\[2em] \dfrac{I\rho_1}{4\pi}(1 + K)\sum\limits_{i=0}^{\infty} K^i\left(\dfrac{1}{\sqrt{\xi^2 + (2iH + h + z)^2}} + \dfrac{1}{\sqrt{\xi^2 + (2iH + h - z)^2}}\right) & z(S) > H,\ z(P) < H \end{cases} \tag{6.7}$$

where $\xi = \sqrt{x^2 + y^2}$.

The infinite series is convergent, although slowly, when $|K| \cong 1$.

As an example, Figure 6.6 shows the soil resistivity at a utility-scale photovoltaic system location, with the soil modeled with reasonable accuracy via a two-layer model.

By applying Eq. 6.7 to the Wenner configuration, under the approximation that $b = a$ (i.e., the pins may be assumed as point sources located on the surface of the ground ($h = 0$)), we obtain:

$$\rho_a = \rho_1\left\{1 + 4\sum_{i=1}^{\infty} K^n\left[\frac{1}{\sqrt{1 + \left(\dfrac{2iH}{a}\right)^2}} - \frac{1}{\sqrt{4 + \left(\dfrac{2iH}{a}\right)^2}}\right]\right\}. \tag{6.8}$$

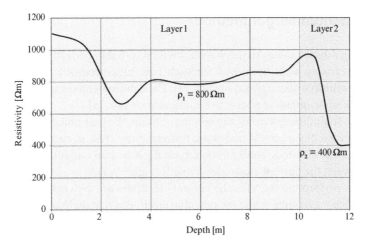

Figure 6.6 Resistivity measured at a utility-scale PV system in Abruzzo region (Italy).

It is worth noting that $\rho_a \rightarrow \rho_1$ when the spacing $a \rightarrow 0$ (i.e., a smaller spacing implies a lower depth of investigation), while $\rho_a \rightarrow \rho_2$ when a is large (i.e., greater spacing implies a higher depth of investigation). From Eq. 6.8, it is evident that the apparent resistivity is not only a function of the spacing a, but also depends on the upper layer depth H, and resistivities and ρ_2.

The mathematical determination of this model's parameters may be challenging due to variations in the structure and properties of the soil. IEEE Std 81 provides iterative methods for an approximate calculation of such parameters.

A popular empirical method for the approximation of the two-layer soil model has been developed by Sunde through a combination of interpolation and field measurements. Sunde proposed a graphical method, based on the interpretation of a series of curves, which are commonly called the "Sunde curves" [11]. IEEE Std 80 describes how to use this empirical approach that allows a first rough estimation of the soil model parameters.

The Wenner method became very popular since its inclusion in the old ASTM Std. G57. It does not require heavy equipment; the results are neither greatly affected by the resistance of the test pins, nor by the holes created by driving the test pins into the soil. Small rods are used as pins, because the theoretical expression of their ground resistance is well known, and because of the easy verification of their burial depth into the soil. A disadvantage is that the rods might vibrate as they are driven into the soil, resulting in poor contact with the earth, thus, making the inversion process more challenging.

6.3 Schlumberger Method

The major disadvantage of the Wenner method is that to perform the CVES all four electrodes must be moved for each new measurement, and this may require multiple operators on site [12, 13]. The Schlumberger arrangement can be a less labor-intensive alternative configuration [14, 15].

In the Schlumberger array, the spacing between the four probes is not equal: the potential electrodes M and N are installed very close to each other at the center of the electrode configuration, typically less than one-fifth of the spacing between the current electrodes A and B (Figure 6.7).

When performing the VES, the current electrodes are systematically moved away from each other, while the potential electrodes remain in place, until the voltage readings become too small for an accurate measurement.

The Schlumberger method provides a better resolution, and the measurements only require the repeated placement of the two outer electrodes. However, due to the larger distance between current and voltage electrodes, this technique requires a voltmeter of greater sensitivity, and higher measurement currents.

Other techniques for soil resistivity measurements are reported in Table 6.2 [16–19].

6.4 Multi-layer Soils

The objective of the soil modeling is to provide a sufficiently accurate approximation of the actual soil conditions and allow the proper grounding system design. When the two-layer model proves to be inadequate to properly describe a more complex soil structures, with the risk of undersizing the grounding system, the three-layer soil model may be adopted. However, three-layer soil

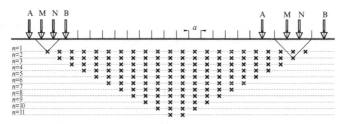

Figure 6.7 Establishment of a 2D electrical resistivity pseudo-section with the Schlumberger configuration.

Table 6.2 Some of the most common used electrode configurations. The letters A and B denote the current electrodes, and the letters M and N denote the potential electrodes.

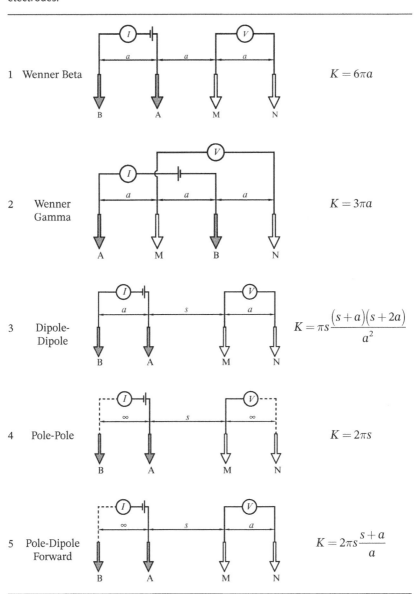

1	Wenner Beta	$K = 6\pi a$
2	Wenner Gamma	$K = 3\pi a$
3	Dipole-Dipole	$K = \pi s \dfrac{(s+a)(s+2a)}{a^2}$
4	Pole-Pole	$K = 2\pi s$
5	Pole-Dipole Forward	$K = 2\pi s \dfrac{s+a}{a}$

(*Continued*)

Table 6.2 (Cont.)

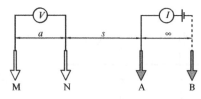

6   Pole-Dipole
Reversed

$$K = 2\pi s \frac{s+a}{a}$$

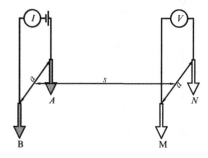

7   Equatorial
Dipole-
Dipole

$$K = \pi s \frac{\sqrt{a^2 + s^2}}{\sqrt{a^2 + s^2} - s}$$

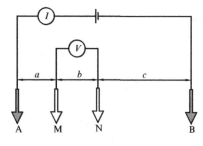

8   Gradient
Array

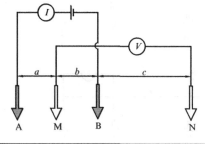

9   Gamma
Array

Table 6.2 (Cont.)

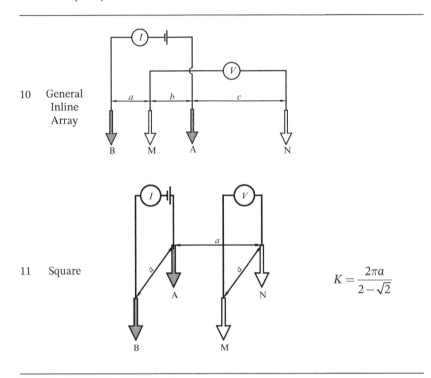

| 10 | General Inline Array | | |
| 11 | Square | $K = \dfrac{2\pi a}{2 - \sqrt{2}}$ | |

models do not add extra accuracy to the calculation of touch and step voltages, which mostly depend on the first layer of the soil.

Two-layer soils [20–22] impact the design of the grounding system resistance, especially when the layers have significantly different resistivities. For multi-layer soils with significant resistivity variations, analytical formulae should be used with caution, and a numerical approach should be adopted.

6.4.1 Ground Grid in Multi-layer Soil

Following is the analysis of the electrical parameters of a ground grid in a multi-layer soil.

For positive values of $K = \dfrac{\rho_2 - \rho_1}{\rho_2 + \rho_1}$ (i.e., $\rho_2 > \rho_1$), as the ground grid burial depth increases, the ground resistance and the mesh voltage decrease at first, and then, approaching the boundary surface between the two layers, they slowly start increasing. In correspondence with the boundary surface, both ground resistance and mesh voltage abruptly increase.

If the burial depth of the ground grid increases, its ground resistance decreases, whereas the mesh voltage increases due to the presence of the higher resistivity layer. For negative values of K, the phenomenon is reversed.

The ground resistance R_G and the mesh voltage V_M as a function of the burial depth of a 20×20 m ground grid, with four meshes, in a two-layer soil ($H = 4$ m) with $\rho_1 = 200 \, \Omega$m and $\rho_2 = 100 \, \Omega$m, are represented in Figure 6.8.

It should be noted that the mesh voltage V_M always tend to increase in the bottom layer as the depth increases, regardless of the sign of K.

For a two-layer soil with a top layer of higher resistivity ($\rho_2 < \rho_1$), an increase in the ground grid burial depth reduces the mesh voltage, the step voltage, the

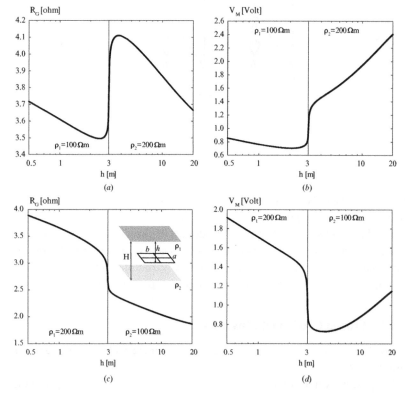

Figure 6.8 Ground resistance and mesh voltage of a groundi grid with four meshes ($a = b = 20$m), burial depth h in a two-layer soil: $h = 3$m; $\rho_1 = 100 \, \Omega$m and $\rho_2 = 200 \, \Omega$m in (a) and (b); $\rho_1 = 200 \, \Omega$m and $\rho_2 = 100 \, \Omega$m in (c) and (d).

ground resistance, and the GPR, which benefits the safety of the persons. In this case, as soon as the ground grid depth reaches the boundary surface between the two soil layers, a sudden decrease in the mesh voltage, step voltage, ground resistance, and GPR occurs.

When the ground grid is buried at a depth greater than the equivalent radius of the grid, its ground resistance becomes almost constant. The increase in the number of the meshes of the ground grid, together with the use of ground rods, flatten the ground potential distribution, and decreases both the ground grid resistance and the ground potential.

6.4.2 Ground Rod in Multi-layer Soil

The behavior of a ground rod in a two-layer soil is shown in Figure 6.9.

The two-layer soil is characterized by $H = 1.5$ m, $\rho_1 = 400\,\Omega$m and $\rho_2 = 50\,\Omega$m; the values of the soil resistivities have also been swapped for a complete analysis.

Two cases have been studied: a ground rod with length $L = 1$ m (i.e., the rod is within the top layer) and $L = 3$ (i.e., the rod penetrates into the bottom layer).

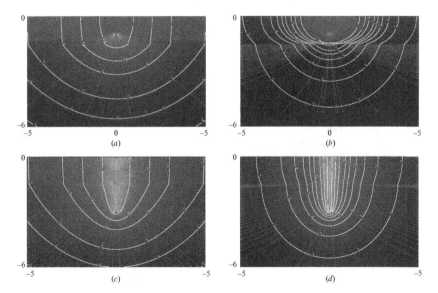

Figure 6.9 Ground rod in two-layer soil ($\rho_1 = 400\,\Omega$m, $\rho_2 = 50\,\Omega$m). The rod has a variable length L = 1 m and L = 3 m. Upper layer at higher resistivity (b) (d); upper layer at lower resistivity (a) (c).

The results clearly show the impact that the two-layer soil has on the current field inside the earth. The layer with the lower resistivity "shunts" the current, allowing a higher density of current lines. In the case of 1 m long rod, when the upper layer has the lower resistivity $\rho_1 = 50\ \Omega\text{m}$, the grounding resistance R_G is $46\ \Omega$, whereas $R_G = 272\ \Omega$ when the resistivity is $\rho_1 = 400\ \Omega\text{m}$; this shows that the ground resistance does not linearly scale up or down with the soil resistivity.

When the rod penetrates the bottom layer, the current's behavior is not uniform along the rod's length. Again, we observe how the current lines are denser within the layer with lower resistivity. Interestingly, most of the current is leaked by the upper section of the rod. In fact, when $\rho_1 = 50\ \Omega\text{m}$ the grounding resistance R_G is $28\ \Omega$, whereas when $\rho_1 = 400\ \Omega\text{m}$ the grounding resistance R_G is $36\ \Omega$. The IEEE Std. 142-2007 indicates that in the first 0.03 m away from the rod surface, 25% of the total resistance occurs.

6.5 Fall-of-Potential Method for Ground Resistance Measurement

When a ground system has been installed, its ground resistance must be measured to ascertain compliance with the design value. The most commonly used method is the 3-point measuring technique, also referred to as the *Fall-of-Potential Method* (FOP) (Figure 6.10) [23–27].

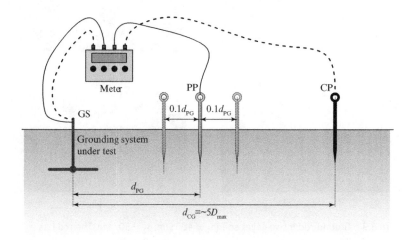

Figure 6.10 Fall-of-potential method.

A current I is impressed across the grounding system (GS) under test and a current probe (CP); the voltage V between the GS and a potential probe (PP) is then measured. The test probes normally offer a higher ground resistance than that of the GS, and must be electrically independent of it. The probe PP is systematically moved along several locations between GS and CP, and the potential difference V across GS and PP is then measured each time; the Ohm's law is used to determine $R_g = V / I$.

To minimize inter-electrode interferences due to mutual resistances, the current probe must be placed at a substantial distance d_{CG} from the ground electrode under test, at least five times the largest dimension D_{max} of the GS under test.

It should be observed that the measurement of grounding resistance is as much an art as it is a science. In fact, the diffculties in determining the correct location of the probes make the fall-of-potential method inconvenient for large grounding systems. For practical fieldwork, the operator's work experience is an essential ingredient.

The distance d_{CG} may make challenging the application of the method to the ground systems of large renewable energy plants, such as utility-scale PV systems, interconnected with wind turbines and/or large HV/MV substations. In those cases, the current probe should be placed at impractical distances away from the GS (e.g., kilometers). The fall-of-potential method would require very long wires for the test leads, and available space to operate (Figure 6.11), and is therefore almost impossible to be applied in urban areas.

An additional difficulty is the determination of the correct location of the PP between GS and CP, i.e., the proper distance d_{PG} with respect to d_{CG} (Figure 6.10).

If PP is placed too close to the GS, the potential probe will be within its area of influence (Figure 6.12 (a)) and as the PP is moved, a steep variation in the reading of the ground resistance will occur [28, 29].

If the CP is correctly positioned, e.g., far away from the GS, a nearly flat ground resistance area somewhere between the CP and the GS will be present, and variations in the position of the PP will produce small changes in the value of the ground resistance. When adequate spacing between electrodes exists, an almost constant value of ground resistance will be read (Figure 6.12 (b)) as the PP is moved [30, 31].

After the first measurement, the PP has moved closer to or away from the GS, by 10% of the original distance d_{PG} (Figure 6.10); the values of the ground resistance are recorded, and PP is again moved until at least three readings are reasonably close to each other.

While performing these measurements, we must always bear in mind that the accuracy of the readings can be affected by buried metal objects near the test probes. This situation may not be encountered in rural and suburban

Figure 6.11 Ground resistance measurements of utility-scale PV plants.

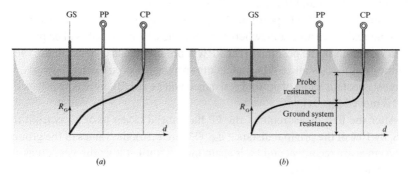

Figure 6.12 Overlapping and non-overlapping shells of grounding resistance.

environments, whereas it is a rather common situation in built environments, where building foundations, buried metal pipes and similar conductive objects are usually present.

The potential probe is typically moved toward the current probe, but it may also be placed in the opposite direction. A rule suggested by IEEE Std. 81 for medium-size GSs is to position the PP such as $d_{PG} = 0.62d_{CG}$, when current and potential probes are aligned. This is rule is referred to as the *62% Method*.

The test current employed in the FOP will also cause a ground potential rise at the current probe, which is typically greater than the ground potential rise of the electrode under test. Thus, the current probe should be inaccessible to persons during the measurement for safety reasons.

6.6 Slope Method for Grounding Resistance Measurement

For large grounding systems, several methods are present in technical literature and standards. One method is the *slope method* [32]. This method requires at least three measurements, which provide three readings R_1, R_2 and R_3, obtained by placing, respectively, the PP at a distance d_{PG} equal to 0.2, 0.4 and 0.6 times the distance d_{CG}. Additional measurements should also be performed to construct a partial graph of the ground resistance R_G as a function of d_{PG}, which would help rule out bad readings, such as abnormal high and low values that could make the subsequent calculation of the ground resistance unreliable, or even provide negative values.

The slope coefficient $\mu = \dfrac{R_3 - R_2}{R_2 - R_1}$ can be calculated to show the rate of change of the ground resistance with the distance. From the slope coefficient, by means of a table included in IEEE Std. 81, we can determine the ratio $d_{PG,opt} / d_{CG}$, where $d_{PG,opt}$ is the optimum distance of the PP at which the *true* ground resistance can be measured.

This method fails when the CP is placed too close to the GS, and a useful indication of this occurrence is apparent when μ cannot be found in the aforementioned table. If this happens, the CP must be further moved away, and the measurements repeated. In addition, to eliminate abnormal readings, it is preferable to test in more than one direction.

It should be, however, noted that slope method is strictly applicable to an N-layer soil, whereas the table available in Std. 81 is for a single-layer soil.

6.7   Star-delta Method for Grounding Resistance Measurement

The *star-delta method* is named after the configuration of the test probes. It is a space-saver configuration, implementing a tighter arrangement of the three probes around the ground electrode under investigation (Figure 6.13). The star-delta method is more suitable for a *point* ground, such as a single rod, rather than for a large grid.

If we denote the unknown grounding resistance as R_1 and the ground resistance of the three probes as R_2, R_3 and R_4, we can express the ground resistance among the probes as the summation of their individual resistances. The method requires three measurements between all pairs of probes (i.e., R_{23}, R_{34} and R_{42}) and three measurements between the probes and the electrode under test located in P_1 (i.e., R_{12}, R_{13} and R_{14}).

The relationships among the grounding resistances are:

$$R_1 = \begin{cases} \dfrac{1}{3}\left[(R_{12}+R_{13}+R_{14}) - \dfrac{(R_{23}+R_{34}+R_{42})}{2}\right] \\[2mm] \dfrac{1}{2}(R_{12}+R_{13}-R_{23}) \\[2mm] \dfrac{1}{2}(R_{12}+R_{14}-R_{24}) \\[2mm] \dfrac{1}{2}(R_{13}+R_{14}-R_{34}) \end{cases} \qquad (6.9)$$

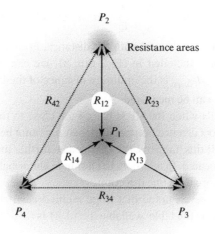

Figure 6.13   Star-delta configuration.

The measurements are considered acceptable if their values match the results of the above systems of equations. If the measurements do not match the results of Eq. 6.9, for instance, because one probe has been placed improperly and its area of influence overlaps with that of the GS, or of another probe, the test must be redone.

It is possible to determine which probe has been misplaced with the following equations, by looking for the result with a match.

$$R_2 = \frac{1}{2}\left(R_{12} + R_{23} - R_{13}\right) = \frac{1}{2}\left(R_{12} + R_{42} - R_{14}\right) = \frac{1}{2}\left(R_{23} + R_{42} - R_{34}\right)$$

$$R_3 = \frac{1}{2}\left(R_{13} + R_{23} - R_{12}\right) = \frac{1}{2}\left(R_{13} + R_{34} - R_{14}\right) = \frac{1}{2}\left(R_{23} + R_{34} - R_{42}\right) \qquad (6.10)$$

$$R_4 = \frac{1}{2}\left(R_{14} + R_{42} - R_{12}\right) = \frac{1}{2}\left(R_{14} + R_{34} - R_{13}\right) = \frac{1}{2}\left(R_{42} + R_{34} - R_{23}\right)$$

6.8 Four Potential Method for Grounding Resistance Measurement

The *Four Potential Method* [33] is based on the fall-of-potential method but allows the user to overcome the problems posed by a complex GS, where the electrical center of the ground grid is difficult to locate.

The theory behind this method is based on ground resistance values obtained by measurements at six different positions of the PP, linked together in four formulae, including the *true* resistance, which would occur with CP at an infinite distance.

The method requires six measurements with the PP at $0.2d_{CP}$, $0.4d_{CP}$, $0.5d_{CP}$, $0.6d_{CP}$, $0.7d_{CP}$ and $0.8d_{CP}$. The true grounding resistance con be determined with the following four formulae:

$$R = \begin{cases} -0.1187R_1 - 0.4667R_2 + 1.9786R_4 - 0.9361R_6 \\ -2.6108R_2 + 4.0508R_3 - 0.1626R_4 - 0.2774R_6 \\ -1.8871R_2 + 1.1148R_3 + 3.6837R_4 - 1.9114R_5 \\ -6.5225R_3 + 13.6816R_4 - 6.8803R_5 + 0.7210R_6 \end{cases} \qquad (6.11)$$

It is expected that the four calculated values of resistances will essentially agree and can then be averaged to provide the final measurement result. If the result of the first approximation is not in agreement with the other three, it may be neglected in the average. Otherwise, if there is a common disagreement, the test measurement must be redone.

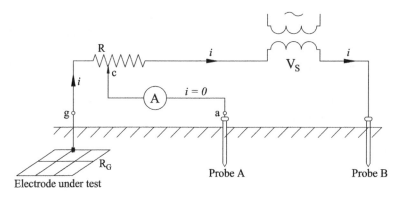

Figure 6.14 Measuring R_G with the potentiometer method.

6.9 Potentiometer Method for Grounding Resistance Measurement

The cursor c of the high-precision potentiometer in Figure 6.14 is operated until a value R of resistance is found when the zero-indicator tester A provides a zero-current reading (or close to zero).

In the above conditions, no current circulates through the probe A, and therefore its ground resistance does not affect the measurement. Thus, $V_{ga} = Ri = R_G i$, and $R = R_G$.

The resistance of the lead to the rheostat may affect the reading, if very long; however, this resistance can be measured and taken into account.

To determine the correct value for R_G, the probe A must be moved alternatively away and toward the probe, until consecutive readings of R on the potentiometer do not appreciably differ.

References

1 Dawalibi, F. and Barbeito, N. (1991). Measurements and computations of the performance of grounding systems buried in multilayer soils. *IEEE Transactions on Power Delivery* 6 (4): 1483–1490.

2 Tabbagh, A., Dabas, M., Hesse, A., and Panissod, C. (2000). Soil resistivity: A non-invasive tool to map soil structure horizonation. *Geoderma* 97 (3): 393–404.

3 Dawalibi, F. and Blattner, C.J. (February 1984). Earth resistivity measurement interpretation techniques. *IEEE Transactions on Power Apparatus and Systems* PAS-103 (2): 374–382.

4 Samouëlian, A., Cousin, I., Tabbagh, A., Bruand, A., and Richard, G. (2005). Electrical resistivity survey in soil science: A review. *Soil and Tillage Research* 83 (2): 173–193.

5 Takahashi, T. and Kawase, T. (1990). Analysis of apparent resistivity in a multi-layer earth structure. *IEEE Transactions on Power Delivery* 5 (2): 604–612.

6 Meliopoulos, A.P. and Papalexpoulos, A.D. (1986). Interpretation of soil resistivity measurements: Experience with the model SOMIP. *IEEE Transactions on Power Delivery* 1 (4): 142–151.

7 Blattner, C.J. (1982). Study of driven ground rods and four point soil resistivity tests. *IEEE Transactions on Power Apparatus and Systems* PAS-101 (8): 2837–2850.

8 Sazali, M.S., Wooi, C.L., Arshad, S.N.M., Wong, T.S., Abdul-Malek, Z., and Nabipour-Afrouzi, H. (2020). Study of soil resistivity using Wenner Four Pin Method: Case study. *2020 IEEE International Conference on Power and Energy (PECon)*, 386–391.

9 Pereira, W.R., Soares, M.G., and Neto, L.M. (2016). Horizontal multilayer soil parameter estimation through differential evolution. *IEEE Transactions on Power Delivery* 31 (2): 622–629.

10 Zhang, L., Yuan, J., and Li, Z. (November 1999). The complex image method and its application in numerical simulation of substation grounding grids. *Communications in Numerical Methods in Engineering* 15 (11): 835–839.

11 Gilbert, G., Chow, Y.L., Bouchard, D.E., and Salama, M.M.A. (2010). Soil model determination using asymptotic approximations to Sunde's curves. *IEEE PES T&D 2010*, 1–7.

12 Zohdy, A.A.R. (February 1989). A new method for the automatic interpretation of Schlumberger and Wenner sounding curves. *GEOPHYSICS* 54 (2): 245–253.

13 Vasantrao, B.M., Bhaskarrao, P.J., Mukund, B.A., Baburao, G.R., and Narayan, P.S. (2017). Comparative study of Wenner and Schlumberger electrical resistivity method for groundwater investigation: A case study from Dhule district (M.S.), India. *Applied Water Science* 7 (8): 4321–4340.

14 Ruan, W., Southey, R.D., Fortin, S., and Dawalibi, F.P. (2005). Effective sounding depths for HVDC grounding electrode design: Wenner versus Schlumberger methods. *2005 IEEE/PES Transmission & Distribution Conference & Exposition: Asia and Pacific*, 1–7.

15 Ibrahim, M.A., Li, L., and Wang, P. (2018). The design of 220kV substation grounding grid with difference soil resistivity using Wenner and Schlumberger methods. *2018 China International Conference on Electricity Distribution (CICED)*, 2525–2530.

16 Tsourlos, P.I., Szymanski, J.E., and Dittmer, J.K. (1995). The topographic effect in Earth resistivity arrays: A comparative study. *1995 International Geoscience and Remote Sensing Symposium, IGARSS '95. Quantitative Remote Sensing for Science and Applications*, vol. 1, 30–32.

17 Johnston, R.H., Trofimenkoff, F.N., and Haslett, J.W. (1987). Resistivity response of a homogeneous Earth with a finite-length contained vertical conductor. *IEEE Transactions on Geoscience and Remote Sensing* GE-25 (4): 414–421.

18 Wiese, T., Greenhalgh, S., Zhou, B., Greenhalgh, M., and Marescot, L. (2015). Resistivity inversion in 2-D anisotropic media: Numerical experiments. *Geophysical Journal International* 201 (1): 247–266.

19 Loke, M.H., Wilkinson, P.B., Uhlemann, S.S., Chambers, J.E., and Oxby, L.S. (December 2014). Computation of optimized arrays for 3-D electrical imaging surveys. *Geophysical Journal International* 199 (3): 1751–1764.

20 Rancic, P.D., Stefanovic, L.V., and Djordjevic, D.J. (1992). A new model of the vertical ground rod in two-layer earth. *IEEE Transactions on Magnetics* 28 (2): 1497–1500.

21 Heppe, R.J. (1979). Step potentials and body currents near grounds in two-layer Earth. *IEEE Transactions on Power Apparatus and Systems* PAS-98 (1): 45–59.

22 Lazzara, J. and Barbeito, N. (1990). Simplified two layer model substation ground grid design methodology. *IEEE Transactions on Power Delivery* 5 (4): 1741–1750.

23 Korasli, C. (2005). Ground resistance measurement with alternative fall-of-potential method. *IEEE Transactions on Power Delivery* 20 (2): 1657–1661.

24 Cheng-gang Wang, T. Takasima, Sakuta, T., and Tsubota, Y. (1998). Grounding resistance measurement using fall-of-potential method with potential probe located in opposite direction to the current probe. *IEEE Transactions on Power Delivery* 13 (4): 1128–1135.

25 Southey, R.D., Siahrang, M., Fortin, S., and Dawalibi, F.P. (2015). Using fall-of-potential measurements to improve deep soil resistivity estimates. *IEEE Transactions on Industry Applications* 51 (6): 5023–5029.

26 Ma, J. and Dawalibi, F.P. (2002). Extended analysis of ground impedance measurement using the fall-of-potential method. *IEEE Transactions on Power Delivery* 17 (4): 881–885.

27 Colella, P., Pons, E., Tommasini, R., Silvestre, M.L.D., Sanseverino, E.R., and Zizzo, G. (2019). Fall of potential measurement of the Earth resistance in urban environments: Accuracy evaluation. *IEEE Transactions on Industry Applications* 55 (3): 2337–2346.

28 Parise, G., Martirano, L., Parise, L., Celozzi, S., and Araneo, R. (2015). Simplified conservative testing method of touch and step voltages by multiple auxiliary electrodes at reduced distance. *IEEE Transactions on Industry Applications* 51 (6): 4987–4993.

29 Hoerauf, R. (2014). Considerations in wind farm grounding designs. *IEEE Transactions on Industry Applications* 50 (2): 1348–1355.

30 Telló, M., Gazzana, D.S., Telló, V.B., Pulz, L.T.C., Leborgne, R.C., and Bretas, A.S. (2020). Substation grounding grid diagnosis applying optimization techniques based on measurements and field tests. *IEEE Transactions on Industry Applications* 56 (2): 1190–1196.

31 Dawalibi, F. and Mukhedkar, D. (1974). Ground electrode resistance measurements in non uniform soils. *IEEE Transactions on Power Apparatus and Systems* PAS-93 (1): 109–115.

32 Tagg, G.F. (1972). Resistance of large earth-electrode systems. *Proceedings of the Institution of Electrical Engineers* 119 (2): 269–272.

33 Tagg, G.F. (1964). Measurement of earth-electrode resistance with particular reference to earth-electrode systems covering a large area. *Proceedings of the Institution of Electrical Engineers* 111 (12): 2118–2130.

Appendix 1

Performance of Grounding Systems in Transient Conditions

Consider well the seed that gave you birth:
you were not made to live your lives as brutes,
but to be followers of worth and knowledge.

Dante Alighieri

The behavior of grounding systems at power frequencies is well understood, and detailed procedures for their design are present in IEEE and IEC standards. However, the performance of grounding systems in transient conditions is quite different; in spite of a large amount of technical literature, there is still no consensus on how to apply the current knowledge to the design of grounding systems for a better high-frequency and dynamic performance.

Currently, several computational tools are available for grounding systems, and classical modeling approaches are based on circuit or transmission-line theory. More rigorous electromagnetic models have been introduced in recent

Electrical Safety Engineering of Renewable Energy Systems. First Edition. Rodolfo Araneo and Massimo Mitolo.
© 2022 by The Institute of Electrical and Electronics Engineers, Inc. Published 2022 by John Wiley & Sons, Inc.

years: remarkable examples are circuit theory method [1–4], transmission line theory method [5–12], electromagnetic field theory method [13–19], such as finite difference time domain method (FDTD) [20, 21], finite element method (FEM) [22–24], method of moments (MoM), and hybrid method [25–30]. Among these numerical methods, the hybrid method [31–38] combines the merits of circuit theory and electromagnetic field theory methods, and allows the direct calculation of the distribution of both branch and leakage currents, providing more accurate results than circuit theory and electromagnetic field theory methods taken singularly. The hybrid approach will be herein discussed [39, 40].

1   Grounding System Analysis

The physical problem is shown in Figure 1: the grounding system (e.g., of a wind turbine) is buried below the ground and is subject to a transient current generated by the lightning strike. Unless otherwise indicated, all the quantities are herein represented in terms of their complex phasors, and a time-harmonic variation of the angular frequency ω is assumed. The grounding system, including floating metallic structure, is assumed to be a network made up of a set of interconnected cylindrical thin conductors placed in any position or orientation. The thin-wire condition implies that the radius is much smaller than the length of the conductors, as is the case of actual electrodes. Although the conductors are considered cylindrical, the method may also be applied to conductors of any other shape by finding an equivalent radius. The proposed

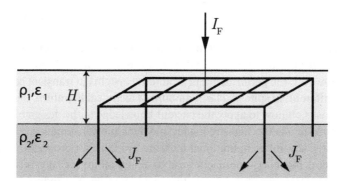

Figure 1   Typical grounding system arrangement: I_F is the fault current flowing into the earthing system and J_F is the fault current density injected into the ground. The ground is assumed stratified and each layer is characterized by the resistivity ρ_i and the permittivity ε_i.

hybrid method aims to obtain an equivalent electric circuit, which takes into account all the mutual inductive, capacitive, and conductive coupling influences among the currents flowing through the conductive segments of the grounding system. Each segment will represent a branch of the equivalent circuit, which will include a serial resistance and self- and mutual inductances.

2   Mathematical Model

In our proposed model, the following assumptions are made:

- the conductors of the grounding system are completely buried in the earth and can be arbitrarily oriented;
- the soil is stratified in N_S planar soil layers;
- each soil layer is characterized by a complex conductivity $\overline{\sigma_i} = \sigma_i + j\omega\varepsilon_0\varepsilon_{r,i}$ (with $i = 1, 2, \ldots, N_S$), where σ_i is the electric conductivity, $\varepsilon_{r,i}$ is the relative dielectric permittivity and ω is the angular frequency of the i-th layer;
- the air is assumed to be a non-conductive medium with permittivity ε_0, and out-of-soil conductors belonging to the grounding system may be present;
- the permeability is assumed equal to μ_0 for all soil layers;
- the constitutive soil parameters can be frequency-dependent;
- non-linear phenomena are neglected.

The grounding system is divided into N_B branches (i.e., segments, see Figure 2), which, in the framework of the network theory, can be studied as elemental oriented segments connecting two nodes (i.e., the terminals of the branches). The discrete grounding system is assumed to have N_N nodes. $I_{LO,k}$ is the longitudinal

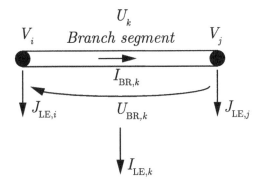

Figure 2   Section of the discretized model: n b $J_{LE,j}$ is the leakage current at node j; $I_{LE,k}$ is the leakage current dispersed by the element k; $I_{BR,k}$ is the current flowing through the element k; V_i is the potential at the node i.

current that flows in the k-th conductor, between the two terminal nodes, whose scalar electric potentials (SEP) V_i are defined with respect to the remote earth, chosen as the zero potential reference.

It is necessary to introduce the nodal voltages because, even though in the low-frequency range the grounding system can be considered equipotential, in transient conditions the voltage drops across the branches may not be negligible due to the electromagnetic coupling and the growth of the internal impedance. In addition, each branch drains a radial leakage current $I_{LE,k}$ to the surrounding earth. The radial current can be assumed uniform if the discretization is sufficiently fine. The grounding system is considered to be energized by the injection of single-frequency sinusoidal source currents F_m at M nodes.

The scalar electric potential along each branch varies between the two values at the terminal nodes. If the minimum wavelength is longer than the segment length, as it is, we can assume that the potential U_k over the k-branch is constant and equal to the average of the voltages V_i and V_j of the two terminal nodes, that is:

$$U_k = \frac{V_i + V_j}{2}. \tag{1}$$

By applying equation 1, the relationship between the $N_B \times 1$ column vector $\mathbf{U}$ of the average SEP of the segments and the $N_N \times 1$ column vector $\mathbf{V}_n$ of the SEP at the nodes can be expressed as:

$$\mathbf{U} = \bar{\mathbf{K}} \cdot \mathbf{V}, \tag{2}$$

where $\bar{\mathbf{K}}$ is a matrix of dimensions $N_B \times N_N$, whose elements are:

$$k_{ij} = \begin{cases} \dfrac{1}{2} & \text{if the branch} i \text{ is incident to node } j \\ 0 & \text{if the branch} i \text{ is not incident to node } j \end{cases} \tag{3}$$

Looking at the leakage currents and the average SEP of all branches, a matricial relationship based on the application of Galerkin's MoM can be written:

$$\mathbf{I}_{LE} = \bar{\mathbf{Y}}_{GC} \cdot \mathbf{U}, \tag{4}$$

where $\mathbf{I}_{LE}$ is the $N_B \times 1$ column vector of the leakage currents and $\bar{\mathbf{Y}}_{GC}$ is a $N_B \times N_B$ square admittance matrix accounting for all the conductive and capacitive couplings among the discretized segments.

Under the assumption that the segmentation is fine with respect to the wavelength, we can divide each leakage current $I_{LE,k}$ into two equal contributions $\dfrac{I_{LE,k}}{2}$ concentrated at the terminal nodes. By this assumption, we obtain a relationship between the concentrated current $\mathbf{J}_{LE}$ at the nodes and the leakage currents $\mathbf{I}_{LE}$, assumed uniformly distributed over the branches:

$$\mathbf{J}_{LE} = \overline{\mathbf{K}}^t \cdot \mathbf{I}_{LE} \tag{5}$$

where $\mathbf{J}_{LE}$ is a $1 \times N_N$ column vector.

Under the above assumptions, the equivalent circuit can be studied by using the node analysis, resulting in the following equation:

$$\mathbf{F} - \mathbf{J}_{LE} = (\overline{\mathbf{A}} \cdot \overline{\mathbf{Y}}_{RL} \cdot \overline{\mathbf{A}}^t) \cdot \mathbf{V}, \tag{6}$$

where $\overline{\mathbf{Y}}_{RL}$ is a $N_B \times N_B$ square admittance matrix accounting for the internal impedance of the branches and the magnetic coupling among them, obtained through the Galerkin's MoM (based on the Sommerfeld integrals of the spectral domain magnetic vector Green's function in stratified media), $\overline{\mathbf{A}}$ is $N_N \times N_B$ branch-to-node incidence matrix whose elements are defined as

$$k_{ij} = \begin{cases} +1 & \text{if the branch } i \text{ enters into node } j \\ -1 & \text{if the branch } i \text{ exits form node } j \\ 0 & \text{if the branch } i \text{ is not incident on node } j \end{cases} \tag{7}$$

and $\mathbf{F}$ is the $N_N \times 1$ column vector of the impressed currents at the nodes that are employed to energize the network.

Substituting all the previous equations into (6), we obtain the final equation:

$$\mathbf{F} = (\overline{\mathbf{K}}^t \cdot \overline{\mathbf{Y}}_{GC} \cdot \overline{\mathbf{K}} + \overline{\mathbf{A}} \cdot \overline{\mathbf{Y}}_{RL} \cdot \overline{\mathbf{A}}^t) \cdot \mathbf{V}. \tag{8}$$

Once the nodal voltages $\mathbf{V}$ are obtained, all the other relevant quantities, that is, branch currents, leakage currents, and surface ground potentials can be straightforwardly determined.

3 Computation of Impedances

In our model, each branch is characterized by lumped self- and mutual impedances, which account for the inductive and capacitive couplings with the other branches. The admittance matrix $\overline{\mathbf{Y}}_{RL}$ is computed as the inverse of the impedance matrix $\overline{\mathbf{Z}}_{RL}$, which accounts for the resistive-inductive couplings, whose elements are obtained as

$$Z_{RL,ij} = \begin{cases} Z_{ii} + j\omega L_{ii} & i = j \\ j\omega L_{ij} & i \neq j \end{cases}, \tag{9}$$

where Z_{ii} is the internal impedance of the i-th cylindrical conductor with radius r_0 at the frequency ω, which is given by the exact formula derived by Schelkunoff in 1934:

$$Z_{ii} = \frac{1}{2\pi r_0} Z_\sigma \frac{I_0(\gamma_\sigma r_0)}{I_1(\gamma_\sigma r_0)}, \tag{10}$$

where $\gamma_\sigma = \sqrt{j\omega\mu_c(\sigma_c + j\omega\varepsilon_c)}$ is the propagation constant in the conducting

material, $Z_\sigma = \dfrac{j\omega\mu_c}{\gamma_\sigma} = \sqrt{\dfrac{j\omega\mu_c}{\sigma_c + j\omega\varepsilon_c}}$ is the wave impedance in the conductor,

and $I_0(\cdot)$ and $I_1(\cdot)$ are modified Bessel functions of the first kind of order 0 and 1, respectively.

In 1973, Wedepohl and Wilcox [41] proposed the following approximation for

the conductor internal impedance $Z_{ij} = \dfrac{Z_\sigma}{2\pi r_0} \coth(0.777 Z_\sigma r_0) + \dfrac{0.3565\, Z_\sigma}{\pi r_0^2}$.

For low frequencies $|\gamma_\sigma r_0| \ll 1$, using the series expansion of the Bessel

functions the internal impedance is $Z_{ij} = R_{dc}\left[1 + \dfrac{1}{48}\left(\dfrac{r_0}{\delta}\right)^2\right] + j\dfrac{\omega\mu_c}{8\pi}$,

where $\delta = \sqrt{\dfrac{2}{\omega\mu_c\sigma_c}}$ is the skin depth in the conductor and $R_{dc} = \dfrac{1}{\pi r_0^2 \sigma_c}$ is the

direct current resistance.

At very high frequencies $|\gamma_\sigma r_0| \gg 1$ and using the asymptotic expressions of

the Bessel functions, the internal impedance becomes $Z_{ij} = \dfrac{1}{2\pi r_0 \sigma_c p}$,

where $p = \sqrt{\dfrac{1}{j\omega\mu_c\sigma_c}}$ is the complex penetration depth for the conductor.

The mutual induction (generally complex and real at low frequency) between conductors i and j is calculated as:

$$L_{ij} = \int_{l_i} dl \int_{l_j} \hat{\mathbf{t}} \cdot \overline{\mathbf{G}}_A(\mathbf{r},\mathbf{r}') \cdot \hat{\mathbf{t}}' dl', \tag{11}$$

where $\overline{\mathbf{G}}_A(\mathbf{r},\mathbf{r}')$ is the dyadic Green function for the magnetic vector potential $\mathbf{A}$ in the layered soil (herein, we use the notation $\mathbf{H} = \dfrac{1}{\mu}\nabla\times\mathbf{A}$), l_i and l_j are the lengths of the i-th observation and j-th source segments with tangents unit vectors $\hat{\mathbf{t}}$ and $\hat{\mathbf{t}}'$, respectively. Further, we indicate with $\mathbf{r}=(\rho,z)$ the observation point and with $\mathbf{r}'=(\rho',z')$ the source point, in cylindrical coordinates.

The admittance matrix $\bar{\mathbf{Y}}_{GC}$ is computed as the inverse of the impedance matrix $\bar{\mathbf{Z}}_{GC}$, which accounts for the conductive-capacitive couplings among the elements. The element $Z_{GC,ij}$ can be calculated by expanding the leakage currents in uniform zero-order basis functions λ_i defined over the i-th branch as $\lambda_i = 1/l_i$ and by applying a standard Galerkin's MoM, as

$$Z_{CG,ij} = \int_{l_j} \frac{dl}{l_i} \int_{l_j} G_\varphi(\mathbf{r},\mathbf{r}') \frac{dl'}{l_j}, \tag{12}$$

where $G_\varphi(\mathbf{r},\mathbf{r}')$ is the scalar potential Green's function in the stratified earth.

4 Green's Function

The hybrid method can be used to study the behavior of any grounding system at d.c., at 50/60 Hz frequency, or under transient discharge via an Inverse Fourier Transform, with a wide-band analysis over the frequency range of the transient current. for this purpose, the correct Green's functions to express the resistive-inductive $\bar{\mathbf{Z}}_{RL}$ and conductive-capacitive $\bar{\mathbf{Z}}_{CG}$ couplings is employed

4.1 Static Formulation

When working at low power frequencies (i.e., 50 or 60 Hz), the static Green's function is sufficiently accurate to calculate mutual inductive and conductive-capacitive couplings. We have determined that this approximation is also accurate to study slow transients.

Let us consider the general case of a layered soil, as depicted in Figure 3: the ground is approximated as an infinite plane, and it is treated as a horizontal N layers, where the conductivity of the m-th layer is σ_m, the dielectric permittivity is ε_m, the depth of each interface is H_m.

Since the ground is usually a non-magnetic material with a permeability μ_0 equal to that of the air, the mutual reactances $j\omega\bar{\mathbf{L}}$ can be calculated by using the uniaxial magnetic Green's function

$$\bar{\mathbf{G}}_A(\mathbf{r},\mathbf{r}') = \begin{pmatrix} 1 & 0 & 0 \\ 0 & 1 & 0 \\ 0 & 0 & 1 \end{pmatrix} \frac{\mu_0}{4\pi} \frac{1}{|\mathbf{r}-\mathbf{r}'|}. \tag{13}$$

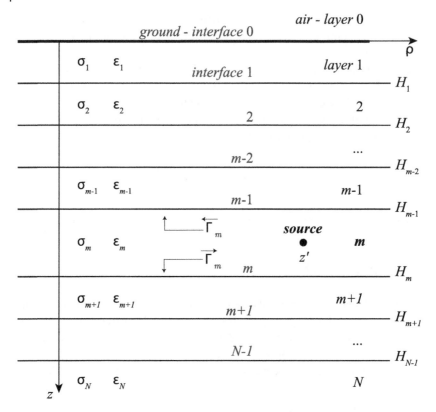

Figure 3 Stratified ground.

As to the conductive and capacitive couplings, i.e., $\overline{\mathbf{Z}}_{GC} = \overline{\mathbf{G}} + j\omega\overline{\mathbf{C}}$, two different problems can be separately solved: one problem for the capacitive couplings is the electrostatic problem of the static electric potential (SEP) V (we use the letter V instead of φ to denote that we are in static conditions) produced by a point charge q_s placed at an arbitrary point z' in an N-layer, assumed insulating, characterized by ε_m; the second problem for the conductive coupling is the electrostatic problem of the SEP associated to the current field produced by a point current source I_s placed at an arbitrary point z' in the N-layer, assumed conductive, characterized by σ_m.

The two problems satisfy a Poisson equation:

$$\nabla^2 V = - \begin{cases} \dfrac{q_s}{\varepsilon}\delta(\mathbf{r}') \\[2ex] \dfrac{I_s}{\sigma}\delta(\mathbf{r}') \end{cases}.$$

(14)

Based on the above, it is possible to compute the Green's function with reference to the problem of the current field through the ground, i.e., $G_{\varphi\sigma}$, and then refer to this solution for the electrostatic problem, i.e., to compute $G_{\varphi\varepsilon}$. In the latter case, it is only necessary to replace the conductivity σ with the dielectric permittivity ε.

In cylindrical coordinates, (14) becomes:

$$\frac{\partial^2 V}{\partial r^2} + \frac{1}{r^2}\frac{\partial V}{\partial r} + \frac{\partial^2 V}{\partial z^2} = -\frac{I_s}{\sigma}\delta(\mathbf{r}'). \tag{15}$$

The boundary conditions are:

$$
\begin{aligned}
\sigma_m \frac{\partial V}{\partial z}\bigg|_{z \to H_m^-} &= \sigma_{m+1}\frac{\partial V}{\partial z}\bigg|_{z \to H_m^+} && \text{current continuity} \\
V\big|_{z \to H_m^-} &= V\big|_{z \to H_m^+} && \text{SEP continuity} \\
\sigma_1 \frac{\partial V}{\partial z}\bigg|_{z \to 0^+} &= 0 && \text{air is a insulator} \\
V\big|_{z,\rho \to \infty} &= 0 && \text{condition at infinity}
\end{aligned}
\tag{16}
$$

By defining the Hankel transform of order 0, we obtain:

$$
\begin{aligned}
\tilde{f}(\lambda) &= \frac{1}{2\pi}\int_0^\infty f(\rho)J_0(\lambda\rho)\rho\,d\rho \\
f(\rho) &= \frac{1}{2\pi}\int_0^\infty \tilde{f}(\lambda)J_0(\lambda\rho)\rho\,d\lambda
\end{aligned}
\tag{17}
$$

where $J_0(\cdot)$ is the Bessel function of order zero, the problem can be recast in the spectral domain as:

$$\frac{d^2\tilde{V}}{dz^2} = \lambda^2\tilde{V} - \frac{I_s}{\sigma}\delta(z - z'). \tag{18}$$

The equation can be rewritten in the following equations:

$$
\begin{aligned}
\frac{d\tilde{V}}{dz} &= -\frac{\lambda^2}{\sigma}\tilde{I} \\
\frac{d\tilde{I}}{dz} &= -\sigma\tilde{V} + \frac{I_s}{\lambda^2}\delta(z - z')
\end{aligned}
\tag{19}
$$

of an equivalent transmission line (TL) with per unit length impedance $Z' = \dfrac{\lambda^2}{\sigma}$ and admittance $Y' = \sigma$; hence, the characteristic impedance is $Z_c = \sqrt{\dfrac{Z'}{Y'}} = \dfrac{\lambda}{\sigma}$ and the propagation constant $\gamma = \sqrt{Z'Y'} = \lambda$. The source is a lumped shunt current generator of value $\dfrac{I_s}{\lambda^2}$.

The TL analogy allows us to treat the problem as a propagation along an equivalent transmission line with uniform sections, each constituted by a ground layer with parameters $Z_{c,m} = \dfrac{\lambda}{\sigma_m}$ and equal propagation constant λ. The transmission line is closed above with an open circuit since $Z_{c,0} = \infty$, and it is closed below on the characteristic impedance of the semi-infinite half-space N, i.e., $Z_{c,N} = \dfrac{\lambda}{\sigma_N}$.

The analogy is very effective. It is possible to calculate the reflection coefficient K_{ij} due to the mismatch between the characteristic impedances of two consecutive TL sections at each interface looking below (between layers m and $m+1$) and above (between layers m and $m-1$):

$$
\begin{aligned}
K_{m,m+1} &= \frac{Z_{c,m+1} - Z_{c,m}}{Z_{c,m+1} + Z_{c,m}} = \frac{\sigma_m - \sigma_{m+1}}{\sigma_m + \sigma_{m+1}} \\
K_{m,m-1} &= \frac{Z_{c,m-1} - Z_{c,m}}{Z_{c,m1} + Z_{c,m}} = \frac{\sigma_m - \sigma_{m-1}}{\sigma_m + \sigma_{m-1}}
\end{aligned}
\tag{20}
$$

Then, it is possible to compute in a recursive manner the voltage *generalized reflection coefficients* at each interface looking below and above with respect to the layer m:

$$
\begin{aligned}
\tilde{\Gamma}_m &= \frac{K_{m,m+1} + \tilde{\Gamma}_{m+1}\, e^{-2\lambda(H_{m+1} - H_m)}}{1 + K_{m,m+1}\tilde{\Gamma}_{m+1}\, e^{-2\lambda(H_{m+1} - H_m)}} \\
\tilde{\Gamma}_m &= \frac{K_{m,m-1} + \tilde{\Gamma}_{m-1}\, e^{-2\lambda(H_{m-1} - H_{m-2})}}{1 + K_{m,m-1}\tilde{\Gamma}_{m-1}\, e^{-2\lambda(H_{m-1} - H_{m-2})}}
\end{aligned}
\tag{21}
$$

Thus, it is possible to study only the m-th layer where the source is placed, bounding the problem at the upper and lower interface through the generalized reflection coefficients. To facilitate the problem solution, we switch to the notation of Chew [42].

Recalling the Lipschitz integral,

$$\frac{1}{\sqrt{\rho^2 + (z - z')^2}} = \int_0^\infty e^{-\lambda|z-z'|} J_0(\lambda\rho) d\lambda, \tag{22}$$

we can express the SEP inside the m-th layer as:

$$V = \frac{I}{4\pi\sigma_m} \int_0^\infty \left[e^{-\lambda|z-z'|} + V_m^+ e^{-\lambda(z-H_{m-1})} + V_m^- e^{+\lambda(z-H_m)} \right] J_0(\lambda\rho) d\lambda, \tag{23}$$

where V_m^+ and V_m^- are the coefficients of forward (downward) and backward (upward) travelling waves along the line reflected at the lower and upper interface of the m-th layer.

Enforcing the boundary conditions:

$$V_m^+ = \bar{\Gamma}_m \left[e^{\lambda(H_{m-1} - z')} + V_m^- e^{+\lambda(H_{m-1} - H_m)} \right] \atop V_m^- = \bar{\Gamma}_m \left[e^{-\lambda(H_m - z')} + V_m^+ e^{-\lambda(H_m - H_{m-1})} \right], \tag{24}$$

and solving for V_m^+ and V_m^-, we obtain:

$$V_m^+ = \bar{\Gamma}_m \frac{e^{-\lambda(z' - H_{m-1})} + e^{-\lambda(H_m - H_{m-1})} \bar{\Gamma}_m e^{+\lambda(z' - H_m)}}{1 - \bar{\Gamma}_m \bar{\Gamma}_m e^{-2\lambda(H_m - H_{m-1})}}$$
$$V_m^- = \bar{\Gamma}_m \frac{e^{\lambda(z' - H_m)} + e^{-\lambda(H_m - H_{m-1})} \bar{\Gamma}_m e^{-\lambda(z' - H_{m-1})}}{1 - \bar{\Gamma}_m \bar{\Gamma}_m e^{-2\lambda(H_m - H_{m-1})}}. \tag{25}$$

When looking at the SEP for a layer $n < m$ (layer n is stacked above layer m), we can express it as:

$$G_{mn} = \frac{1}{4\pi\sigma_m} \int_0^\infty A^- \left[e^{\lambda z} + \bar{\Gamma}_n e^{-\lambda(z - 2H_{n-1})} \right] J_0(\lambda\rho) d\lambda. \tag{26}$$

The magnitude factor A^- can be obtained recursively starting from layer m up to layer n as:

$$A_m^- = \frac{e^{-\lambda z'} + \bar{\Gamma}_m e^{+\lambda(z'-2H_m)}}{1 - \bar{\Gamma}_m \bar{\Gamma}_m e^{-2\lambda(H_m - H_{m-1})}}$$

$$A_i^- = A_{i+1}^- S_{i+1,i}^- = A_{i+1}^- \left[\frac{T_{i+1,i}}{1 - K_{i,i+1} \bar{\Gamma}_i e^{-2\lambda(H_i - H_{i-1})}} \right].$$

(27)

where $T_{ij} = 1 + K_{ij}$ is the transmission coefficient and $K_{ij} = -K_{ji}$.

When looking at the SEP for a layer $n > m$ (layer n is stacked below layer m), we can express it as:

$$G_{mn} = \frac{1}{4\pi\sigma_m} \int_0^\infty A^+ \left[e^{-\lambda z} + \bar{\Gamma}_n e^{\lambda(z - 2H_n)} \right] J_0(\lambda\rho) d\lambda,$$

(28)

and the new amplitude factor A^+ can be obtained recursively starting from layer m down to layer n as:

$$A_m^+ = \frac{e^{\lambda z'} + \bar{\Gamma}_m e^{-\lambda(z'-2H_{m-1})}}{1 - \bar{\Gamma}_m \bar{\Gamma}_m e^{-2\lambda(H_m - H_{m-1})}}$$

$$A_i^+ = A_{i-1}^+ S_{i-1,i}^+ = A_{i-1}^+ \left[\frac{T_{i-1,i}}{1 - K_{i,i-1} \bar{\Gamma}_i e^{-2\lambda(H_i - H_{i-1})}} \right].$$

(29)

Let us now consider some cases of practical importance.

4.1.1 One-Layer Ground

The ground is assumed to be a half-space layer (layer n. 1) placed below the air (layer n. 0.) The electrode source is placed in the ground. The reflection coefficients are $\bar{\Gamma}_1 = K_{10} = 1$ and $\bar{\Gamma}_1 = 0$. The potentials along the two layers are respectively:

$$G_{10} = \frac{1}{2\pi\sigma_1} \int_0^\infty e^{\lambda(z-z')} J_0(\lambda\rho) d\lambda$$

(30)

$$G_{11} = \frac{1}{4\pi\sigma_1} \int_0^\infty \left[e^{-\lambda|z-z'|} + e^{-\lambda(z+z')} \right] J_0(\lambda\rho) d\lambda$$

(31)

By using the image theory, the two fields can be calculated as:

$$G_{10} = \frac{1}{2\pi\sigma_1} \frac{1}{\sqrt{\rho^2 + (z-z')^2}}$$

(32)

$$G_{11} = \frac{1}{4\pi\sigma_1}\left(\frac{1}{\sqrt{\rho^2 + (z - z')^2}} + \frac{1}{\sqrt{\rho^2 + (z + z')^2}}\right) \tag{33}$$

4.1.2 Two-Layer Ground

The ground is assumed to be a horizontally stratified two-layer soil where the first upper layer has depth h.

If the electrode is placed in the top layer, the potentials in the three regions can be expressed as:

$$G_{10} = \frac{1}{2\pi\sigma_1}\int_0^\infty V_0^- e^{\lambda z} J_0(\lambda\rho)d\lambda \tag{34}$$

$$G_{11} = \frac{1}{4\pi\sigma_1}\int_0^\infty \left[e^{-\lambda|z-z'|} + V_1^+ e^{-\lambda z} + V_1^- e^{+\lambda(z-H)}\right] J_0(\lambda\rho)d\lambda \tag{35}$$

$$G_{12} = \frac{1}{4\pi\sigma_1}\int_0^\infty V_2^+ e^{-\lambda z} J_0(\lambda\rho)d\lambda \tag{36}$$

where

$$\begin{aligned}
V_0^- &= 2\frac{e^{-\lambda z'} + e^{+\lambda(z'-2H)}K_{12}}{1 - K_{12}e^{-2\lambda H}} \\
V_1^+ &= \frac{e^{-\lambda z'} + K_{12}e^{-\lambda(z'-2H)}}{1 - K_{12}e^{-2\lambda H}} \quad; \quad V_1^- = K_{12}\frac{e^{-\lambda(z'+H)} + e^{\lambda(z'-H)}}{1 - K_{12}e^{-2\lambda H}} \\
V_2^+ &= \frac{e^{-\lambda z'} + e^{+\lambda z'}}{1 - K_{12}e^{-2\lambda H}}(1 + K_{12})
\end{aligned} \tag{37}$$

By using the image theory, the SEP can be calculated as:

$$G_{10} = \frac{1}{2\pi\sigma_1}\sum_{k=0}^\infty K_{12}^k \left[\frac{1}{\sqrt{\rho^2 + (z - z' - 2kH)^2}} + \frac{K_{12}}{\sqrt{\rho^2 + (z + z' - 2kH)^2}}\right] \tag{38}$$

$$G_{11} = \frac{1}{4\pi\sigma_1} \left\{ \frac{1}{\sqrt{\rho^2 + (z-z')^2}} + \frac{1}{\sqrt{\rho^2 + (z+z')^2}} + \right.$$

$$\sum_{k=1}^{\infty} K_{12}^k \left[\frac{1}{\sqrt{\rho^2 + (z-z'-2kH)^2}} + \frac{1}{\sqrt{\rho^2 + (z+z'+2kH)^2}} \right] + \quad (39)$$

$$\left. \sum_{k=1}^{\infty} K_{12}^k \left[\frac{1}{\sqrt{\rho^2 + (z+z'-2kH)^2}} + \frac{1}{\sqrt{\rho^2 + (z+z'-2kH)^2}} \right] \right\}$$

$$G_{12} = \frac{1}{4\pi\sigma_1} (1 + K_{12}) \sum_{k=0}^{\infty} K_{12}^k \left[\frac{\frac{1}{\sqrt{\rho^2 + (z+z'+2kH)^2}} +}{\frac{1}{\sqrt{\rho^2 + (z-z'+2kH)^2}}} \right] \quad (40)$$

If the electrode is embedded into the lower ground layer, the potentials in the three regions can be expressed as:

$$G_{20} = \frac{1}{2\pi\sigma_2} \int_0^{\infty} V_0^- e^{\lambda z} J_0(\lambda\rho) d\lambda \quad (41)$$

$$G_{21} = \frac{1}{4\pi\sigma_2} \int_0^{\infty} V_1^- \left[e^{\lambda z} + e^{-\lambda z} \right] J_0(\lambda\rho) d\lambda \quad (42)$$

$$G_{22} = \frac{1}{4\pi\sigma_2} \int_0^{\infty} \left[e^{-\lambda|z-z'|} + V_2^+ e^{-\lambda(z-H)} \right] J_0(\lambda\rho) d\lambda \quad (43)$$

Where:

$$V_0^- = 2V_1^-$$

$$V_1^- = (1 - K_{12}) \frac{e^{-\lambda z'}}{1 - K_{12}e^{-2\lambda h}} \quad (44)$$

$$V_2^+ = \frac{e^{-\lambda(z'+H)} - K_{12}e^{-\lambda(z'-H)}}{1 - K_{12}e^{-2\lambda h}}$$

Using the image theory, the SEP can be calculated as:

$$G_{20} = \frac{1}{2\pi\sigma_2}(1-K)\sum_{k=0}^{\infty}\frac{K^i}{\sqrt{\rho^2 + (z-z'-2kH)^2}} \tag{45}$$

$$G_{21} = \frac{1}{4\pi\sigma_2}\left\{(1-K)\sum_{k=0}^{\infty}\left[\frac{\frac{K^i}{\sqrt{\rho^2 + (z-z'-2kH)^2}}+}{\frac{K^i}{\sqrt{\rho^2 + (z+z'-2kH)^2}}}\right]\right\} \tag{46}$$

$$G_{22} = \frac{1}{4\pi\sigma_2}\left|\begin{array}{c}\frac{1}{\sqrt{\rho^2 + (z-z')^2}} - \frac{K}{\sqrt{\rho^2 + (z+z'-2H)^2}}+\\ (1-K^2)\sum_{k=0}^{\infty}\frac{K^i}{\sqrt{\rho^2 + (z+z'+2kH)^2}}\end{array}\right| \tag{47}$$

4.2 Dynamic Formulation

The formulation of the spatial-domain hybrid method requires the calculation of the Green's functions of the vector $\bar{\mathbf{G}}_A(\mathbf{r},\mathbf{r}')$ and scalar $G_\varphi(\mathbf{r},\mathbf{r}')$ potentials in a multilayer medium. Vast literature is available on the definition of Green's functions in layered media, starting from the pioneer works by Michalski and Zheng [43], passing through the milestone work of Michalski and Mosig [44].

Starting from the *formulation C* proposed in [42], the dyadic magnetic vector potential Green's function $\bar{\mathbf{G}}_A(r,r')$ can be defined as:

$$\bar{\mathbf{G}}_A(\mathbf{r},\mathbf{r}') = \begin{bmatrix} \bar{G}_{xx}^A & 0 & 0 \\ 0 & \bar{G}_{yy}^A & 0 \\ \bar{G}_{zx}^A & \bar{G}_{zy}^A & \bar{G}_{zz}^A \end{bmatrix}. \tag{48}$$

where we have dropped the dependence on $\mathbf{r}$ and $\mathbf{r}'$ to keep the notation compact.

Introducing the Fourier-Bessel transforms pair of order n as:

$$\tilde{f}(k_\rho) = \int_0^\infty f(k_\rho)J_n(k_\rho\rho)\rho\,d\rho$$

$$f(\rho) = \int_0^\infty \tilde{f}(k_\rho)J_n(k_\rho\rho)k_\rho\,dk_\rho \tag{49}$$

where $J_n(\cdot)$ is the Bessel function of order n and $k_\rho = \sqrt{k_x^2 + k_y^2}$.

It is possible to define the following Sommerfeld integral - $S_n\{\tilde{f}(k_\rho)\} = \dfrac{1}{2\pi}\displaystyle\int_0^\infty \tilde{f}(k_\rho) J_n(k_\rho\rho) k_\rho \, dk_\rho$ of order n. Consequently, the compo-

nents of $\bar{\mathbf{G}}_A(\mathbf{r},\mathbf{r}')$ can be expressed as

$$
\begin{aligned}
G_{xx}^A(\acute{\mathbf{A}}z\,|\,z') &= G_{yy}^A(\rho,z\,|\,z') = S_0\{\tilde{G}_{vv}(k_\rho,z\,|\,z')\} \\
G_{zz}^A(\acute{\mathbf{A}}z\,|\,z') &= S_0\{\tilde{G}_{zz}(k_\rho,z\,|\,z')\} \\
G_{zx}^A(\acute{\mathbf{A}}z\,|\,z') &= -j\cos\varphi\, S_1\{\tilde{G}_{zu}(k_\rho,z\,|\,z')\} \\
G_{zy}^A(\acute{\mathbf{A}}z\,|\,z') &= -j\sin\varphi\, S_1\{\tilde{G}_{zu}(k_\rho,z\,|\,z')\}.
\end{aligned}
\tag{50}
$$

The spectral domain Green's functions $\tilde{G}_{vv,zz,zu}(k_\rho,z\,|\,z')$ are derived with the transmission line analogy as:

$$
\begin{aligned}
\tilde{G}_{vv}^A(z\,|\,z') &= \frac{1}{j\omega\mu_0} V_i^h \\
\tilde{G}_{zz}^A(z\,|\,z') &= \frac{1}{j\omega\varepsilon_0} \frac{\mu_r}{\bar{\varepsilon}_r'} I_v^e \quad . \\
\tilde{G}_{zu}^A(z\,|\,z') &= \frac{\mu_r}{jk_\rho}\left[I_i^h - I_i^e \right]
\end{aligned}
\tag{51}
$$

(we have dropped the dependence on z and z' to keep the notation compact)

where $\eta_0 = \sqrt{\dfrac{\mu_0}{\varepsilon_0}}$ is the wave impedance in free space, μ_0 and ε_0 are respectively the free space permeability and permittivity, whereas μ_r and $\bar{\varepsilon}_r = \varepsilon_r - j\dfrac{\sigma}{\omega\varepsilon_0}$ are respectively the relative permeability and complex permittivity of the media, where the source or observation point (indicated with the prime symbol) is located. Furthermore:

- V are the modal voltages;
- I are the modal currents;
- the superscript letter indicate the type of transmission line (i.e, e means TM mode and h means TE mode);
- the subscript letter indicates the source type placed at z', i.e., v means a voltage source and i means a current source.

We will illustrate the computation of the modal quantities in the next paragraph.

It is necessary to define the scalar potential Green's function $G_\varphi(\mathbf{r}, \mathbf{r}')$. In a homogeneous media, the scalar potential φ is related to the vector potential $\mathbf{A}$ through the Lorenz gauge $\varphi = \dfrac{1}{-j\omega\mu\bar{\varepsilon}} \nabla \cdot \mathbf{A}$. Consequently, we have

$$\varphi = \frac{j}{\omega\mu\bar{\varepsilon}} \int_{S'} \left[\nabla \cdot \bar{\mathbf{G}}_A(\mathbf{r} \mid \mathbf{r}') \right] \mathbf{J}(\mathbf{r}') dS'$$ and, to deduce the scalar-potential Green's

function, we need to transform the divergence operator to act on the current density in view of the continuity equation $\nabla \cdot \mathbf{J} = -j\omega q$.

It is, therefore, immediate to define $\dfrac{j}{\omega\mu\bar{\varepsilon}} \nabla \cdot \bar{\mathbf{G}}_A(\mathbf{r} \mid \mathbf{r}') = \dfrac{1}{j\omega} \nabla' G_\varphi(\mathbf{r} \mid \mathbf{r}')$. The

problem that we meet at this point is that if the soil is stratified, a scalar electric potential Green's function that fulfills the previous relation does not exist, because it cannot be independent of the orientation of $\mathbf{J}$. The physical reason is that the scalar potential of a point charge associated with vertical or horizontal current dipoles in a layered medium is generally different.

Michalski proposed a gauge [43] with the introduction of a scalar function $K_\varphi(\mathbf{r} \mid \mathbf{r}')$, that resembles the original electric scalar Green's function, plus a

correction term $C_\varphi(\mathbf{r} \mid \mathbf{r}')\hat{\mathbf{z}}$ [44], so that $\dfrac{j}{\omega\mu\bar{\varepsilon}} \nabla \cdot \bar{\mathbf{G}}_A(\mathbf{r} \mid \mathbf{r}') = \dfrac{1}{j\omega}\Big[\nabla' K_\varphi(\mathbf{r} \mid \mathbf{r}') - $

$C_\varphi(\mathbf{r} \mid \mathbf{r}')\hat{\mathbf{z}}\Big]$. According to this gauge, the scalar potential can be expressed as:

$$\varphi = \frac{1}{\varepsilon} \int_{S'} K_\varphi(\mathbf{r} \mid \mathbf{r}') q(\mathbf{r}') dS' - \frac{1}{j\omega\bar{\varepsilon}} \int_{S'} C_\varphi(\mathbf{r} \mid \mathbf{r}') \hat{\mathbf{z}} \cdot \mathbf{J}(\mathbf{r}') dS'. \tag{52}$$

Now the scalar potential depends not only on the point charge source but also on the current source to which the charge is associated. It is apparent that $K_\varphi(\mathbf{r} \mid \mathbf{r}')$ can be interpreted as the scalar potential of a point charge associated with a horizontal dipole, since the correction term acts only on vertical dipoles.

The two terms can be obtained from the spectral counterparts as $K_\varphi(\acute{\mathbf{A}}z \mid z') = S_0\{\tilde{K}_\varphi(k_\rho, z \mid z')\}$ and $C_\varphi(\acute{\mathbf{A}}z \mid z') = S_0\{\tilde{C}_\varphi(k_\rho, z \mid z')\}$, where

$$\tilde{K}_\varphi(z \mid z') = -\frac{j\omega\varepsilon_0}{k_\rho^2}\left[V_i^h - V_i^e\right]$$

$$\tilde{C}_\varphi(z \mid z') = \left(\frac{k_0}{k_\rho}\right)^2 \mu_r'\left[V_v^h - V_v^e\right], \tag{53}$$

and where $k_0 = \omega\sqrt{\mu_0\varepsilon_0}$ is the wavenumber in free space.

This formulation (denoted as *formulation C*) enjoys a clear advantage over other formulations because undesirable contour integrals cancel; therefore, the scalar potential kernel is continuous with respect to z', and $K_\varphi(\mathbf{r}|\mathbf{r}')$ is continuous at the interfaces with respect to z, which results in considerable simplifications in the numerical procedure when the source electrode penetrates one or more interfaces.

Since the formulation is not friendly in a classical *low-frequency way of thought* where the scalar potential is generated only by a point source charge, the formulation can be changed according to the suggestions of Michalski [43] and as done explicitly by Cangellaris in [45], by incorporating the gradient of the correction term into the vector dyadic Green's function to obtain the following expressions:

1) a new magnetic potential dyadic Green's function $\bar{\mathbf{G}}_A(\mathbf{r},\mathbf{r}')$:

$$\bar{\mathbf{G}}_A(\mathbf{r},\mathbf{r}') = \begin{vmatrix} \bar{G}^A_{xx} & 0 & -\dfrac{\mu'_r}{\mu_r}\bar{G}^A_{zx} \\ 0 & \bar{G}^A_{yy} & -\dfrac{\mu'_r}{\mu_r}\bar{G}^A_{zy} \\ \bar{G}^A_{zx} & \bar{G}^A_{zy} & \bar{G}^A_{zz} \end{vmatrix} \tag{54}$$

where (50) still holds but with the following spectral counterparts:

$$\tilde{G}^A_{vv}(z\,|\,z') = \frac{1}{j\omega\mu_0}V^h_i$$

$$\tilde{G}^A_{zz}(z\,|\,z') = \frac{1}{j\omega\varepsilon_0}\left[\left(\frac{\mu_r}{\bar{\varepsilon}_r} + \frac{\mu'_r}{\bar{\varepsilon}_r}\right)I^e_v + \frac{k_0^2\mu_r\mu'_r}{k_\rho^2}\left(I^h_v - I^e_v\right)\right] \tag{55}$$

$$\tilde{G}^A_{zu}(z\,|\,z') = \frac{\mu_r}{jk_\rho}\left[I^h_i - I^e_i\right]$$

2) an electric potential scalar Green's function which coincides with $K_\varphi(z\,|\,z')$. Thus, we can express the electric field in the classical way, i.e.,

$$\mathbf{E} = -j\omega\nabla\cdot\mathbf{A} - \nabla\varphi = -j\omega\mu_0\int_{S'}\left[\nabla\cdot\bar{\mathbf{G}}_A(\mathbf{r}|\mathbf{r}')\right]\mathbf{J}(\mathbf{r}')dS'$$

$$-\nabla\left[\frac{1}{\varepsilon}\int_{S'}K_\varphi(\mathbf{r}|\mathbf{r}')q(\mathbf{r}')dS'\right]. \tag{56}$$

The new kernels $\bar{G}_A(\mathbf{r}\,|\,\mathbf{r}')$ and $\dfrac{K_\varphi(\mathbf{r}\,|\,\mathbf{r}')}{j\omega}$ (the division by $j\omega$ is due to the fact that we derived the dynamic formulation using the complex permittivity and not the complex conductivity as we did in the low-frequency formulation) can be directly used in (11) and (12), respectively, to apply the proposed hybrid circuital-electromagnetic procedure.

The following mathematical treatment should not mislead the reader. We have met two fundamental concepts:

(i) at high frequencies, when the displacement currents cannot be neglected, the electric potential cannot be defined as it can in stationary conditions, where the electric field is conservative, and the Coulomb gauge holds; so, the scalar potential φ is a generalization of the potential V;

(ii) in a stratified medium, a correction term to define the electric potential is needed, since it depends not only on the charge point source, but also on the associated dipole; therefore, in light of the developed formulation, every time a vertical electrode is considered, the radiated potential must account for the correction term (52);

(iii) we have developed an hybrid approach in which $\bar{G}_A(\mathbf{r}\,|\,\mathbf{r}')$ and $K_\varphi(\mathbf{r}\,|\,\mathbf{r}')$ resembles at high frequencies the inductive and capacitive couplings that are correctly modelled in the d.c. formulation; however, the couplings must be considered strictly joined to maintain a physical meaning for the overall formulation, and cannot be separated; it is apparent that the numerical evaluation of the two terms depends on where the correction term is considered.

4.2.1 Equivalent Transmission Line Approach

To obtain the Green's function, we must overcome the problem of the propagation along a cascade of transmission lines (TLs), where each segment n is described, from the electrical point of view, by its propagation constant and characteristic impedance.

In general, these are different for the e- and h-modes (hereafter, we will distinguish the two kinds of modes with the superscript $q = \{e,h\}$): the propagation constant is $k_{z,n} = \sqrt{k_0^2 \mu_{r,n}\bar{\varepsilon}_{r,n} - k_\rho^2}$ and is equal for both the e and h modes,

while the characteristic impedances are $Z_{c,n}^e = \dfrac{k_{z,n}}{\omega\varepsilon_0\bar{\varepsilon}_{r,n}}$ for e-modes and

$Z_{c,n}^h = \dfrac{\omega\mu_0\mu_{r,n}}{k_{z,n}}$ for h-modes. The boundaries of the n-th layer are placed in

H_{n-1} and H_n as reported in Figure 3 and its thickness is $h_n = H_n - H_{n-1}$.

The TL equations for a unit-strength shunt current source placed at z' are

$$\frac{dV_i^q}{dz} = -jk_{z,n}Z_{c,n}^q I_i^q$$
$$\frac{dI_i^q}{dz} = -jk_{z,n}\frac{1}{Z_{c,n}^q}V_i^q + \delta(z - z'),$$

(57)

where V_i^q and I_i^q denote the voltage and current at z, respectively, due to a 1-A shunt current source. Similarly, the TL equations for a unit-strength series voltage source placed at z' are

$$\frac{dV_v^q}{dz} = -jk_{z,n}Z_{c,n}^q I_v^q + \delta(z - z')$$
$$\frac{dI_v^q}{dz} = -jk_{z,n}\frac{1}{Z_{c,n}^q}V_v^q,$$

(58)

where V_v^q and I_v^q denote the voltage and current at z, respectively, due to a 1-V series voltage source.

From the previous equations, it should be observed that four problems must be solved to determine eight variables, i.e., the voltages and currents at the observation point z due to a voltage or a current source placed at z' along TM (e) or TE (h) equivalent TLs. However, the problem can be written in a compact form to deal simultaneously with voltages and currents independently of the type of modes (since the change of the mode implies only the change of the characteristic impedance) and independently of the source generator (since it implies only a correction term on the source).

To this end, we define $P = \{V, I\}$ and we define the general quantity $P_{P'}^q(z \mid z')$, where the subscript letter indicates the type of source. We then apply the reciprocity theorem to limit our analysis only to the case where the observation point is placed below the source point, i.e., $z \geq z'$ (the reader should note the z-axis points oriented toward the depth of the soil).

The solutions to (57) and (58) satisfy the following reciprocity relations:

$$V_i^q(z \mid z') = V_i^q(z' \mid z)$$
$$I_v^q(z \mid z') = I_v^q(z' \mid z)$$
$$V_v^q(z \mid z') = -I_i^q(z' \mid z)$$
$$I_i^q(z \mid z') = -V_v^q(z' \mid z)$$

(59)

To express a closed-form solution of the inhomogeneous equation in a TL section, we introduce the following variables:

$$\sigma = \begin{cases} +1 & P = V \\ -1 & P = I \end{cases}, \sigma' = \begin{cases} +1 & P' = V \\ -1 & P' = I \end{cases} \quad S = \begin{cases} +1 & P = V \\ +\dfrac{1}{Z} & P = I \end{cases}$$

$$s' = \frac{1}{2} \begin{cases} U & P' = V \\ Z'J & P' = I \end{cases}, \tag{60}$$

where we have denoted the characteristic impedance of the layers where the source and observation points are placed as Z' and Z, respectively, for conciseness. In the definition of s', $U = 1$ is the amplitude of the series voltage generator and $J = 1$ is the amplitude of the shunt current generator.

Then, we need to calculate the *generalized reflection coefficients* at each interface looking below ($\vec{\Gamma}_i$) and above ($\overleftarrow{\Gamma}_i$) with respect to the i-th layer, as already done in (21) for the static case.

To this purpose, we recall the Fresnel reflection coefficient $R_{i,j} = \dfrac{Z_j - Z_i}{Z_j + Z_i}$,

that is now defined through the characteristic impedance of the dynamic TM or TE modes in each layer. We can compute $\vec{\Gamma}_i$ recursively starting from the last layer N and going backward to the first one as:

$$\vec{\Gamma}_i = \frac{R_{i,i+1} + \vec{\Gamma}_{i+1} t_{i+1}}{1 + R_{i,i+1} \vec{\Gamma}_{i+1} t_{i+1}}, \tag{61}$$

where $t_{i+1} = \exp\left(-j2k_{z,i+1} h_{i+1}\right)$ and $\vec{\Gamma}_{N+1} = 0$. Similarly, we can compute $\overleftarrow{\Gamma}_i$ proceeding from the first layer to the last one as:

$$\overleftarrow{\Gamma}_i = \frac{R_{i,i-1} + \overleftarrow{\Gamma}_{i-1} t_{i-1}}{1 + R_{i,i-1} \overleftarrow{\Gamma}_{i-1} t_{i-1}}, \tag{62}$$

where $t_{i-1} = \exp\left(-j2k_{z,i-1} h_{i-1}\right)$ and $\overleftarrow{\Gamma}_0 = 0$.

Furthermore, we define the *generalized transmission coefficient* $\vec{\tau}_i$ looking below as:

$$\vec{\tau}_i = \frac{1 - R_{i-1,i} \vec{\Gamma}_{i-1}}{1 - R_{i-1,i}}, \tag{63}$$

with $\vec{\tau}_i = 0$ (note that if $R_{i-1,i} = 1$, then $\vec{\tau}_i = 0$). We do not need the transmission coefficient looking above $\overleftarrow{\tau}_i$ since we have restricted our analysis to the case $z \geq z'$, thanks to the reciprocity.

According to these positions, the *intra-layer interaction* in the i-th layer can be expressed as

$$P_{p'}^q(z \mid z') = ss'\left(e^{-jk_{z,i}|z-z'|} + \frac{1}{D}\sum_{k=1}^{4}a_k e^{-jk_{z,i}d_k}\right),$$

(64)

where:

$$
\begin{array}{ll}
a_1 = \sigma\bar{\Gamma}_i & d_1 = 2H_i - |z+z'| \\
a_2 = -\sigma'\bar{\Gamma}_i & d_2 = |z+z'| - 2H_{i-1} \\
a_3 = \bar{\Gamma}_i\bar{\Gamma}_i & d_3 = 2H_{i-1} + |z-z'| \\
a_4 = a_1 a_2 & d_4 = 2H_{i-1} - |z-z'|
\end{array}
$$

(65)

and $D = 1 + \bar{\Gamma}_i\bar{\Gamma}_i e^{-j2k_{z,i}h_i}$.

The *inter-layer interaction* between a source placed in the layer i and an observation point placed in the layer j, reads:

$$P_{p'}^q(z \mid z') = \frac{ss'}{D}\left[1 - \sigma'\bar{\Gamma}_i e^{-j2k_{z,i}(z'-H_{i-1})}\right]e^{-jk_{z,i}(H_i-z')}\vec{T}_{ij}\,e^{-jk_{z,i}(z-H_{j-1})}$$
$$\left[1 - \sigma\bar{\Gamma}_j e^{-j2k_{z,j}(H_j-z')}\right],$$

(66)

where $\vec{T}_{ij}$ is the generalized transmission coefficient through layers defined as:

$$\vec{T}_{ij} = \vec{\tau}_j \prod_{k=i+1}^{j-1}\vec{\tau}_k e^{-jk_{z,k}h_k}.$$

(67)

5 Numerical Integration Aspects

The solution of the proposed procedure requires the treatment of some numerical problematic issues related to the singularity that arise in the spatial domain and to the integration of the oscillatory Sommerfeld integrals in the spectral domain related to the Hankel transforms defined in (17) and the Fourier-Bessel transforms defined in (49).

5.1 Singular Term

The evaluation of the admittance and impedance matrices requires the computation of each linear wire segment that is part of the grounding system. Since the functions for currents and charges are constant and the vector bases for current have constant direction on each linear segment, when the observation j and source i segments coincide, both the integrals in (11) and (12) reduce to a scalar integral I of the form:

$$I = c_\alpha \int_{l_j} dl \int_{l_i} K(\mathbf{r},\mathbf{r}') dl'. \tag{68}$$

where c_α is a suitable coefficient and $K(\mathbf{r},\mathbf{r}')$ is the corresponding kernel (e.g., the component of the dyadic vector Green's function or the scalar Green's function in the space domain). As previously discussed, the Green's function in the spectral domain within the soil layer n is always composed by a first term that represents the direct ray between the source and the observation point and a second term that represents the superposition of the rays that undergo partial reflections at the upper and lower slab boundaries before reaching the field point. The first term, when is transformed back from the spectral to the spatial domain, contains a singular term of the form $\frac{1}{R}$ and that must be accurately treated when the proposed analysis is applied to wire structures, so that to prevent non-essential (i.e., integrable) singularities. Following the procedure presented in [46, 47], the sources and the potentials about a linear tubular section present a rotational symmetry (hereafter, we refer to a cylindrical coordinate system (ρ,ϕ,z)). Assuming no loss of generality, for sources distributed uniformly on a cylindrical tube of constant radius r_0 centered along the z-axis and an observation point with cylindrical coordinates $(r_0,0,z)$, the wire kernel $K(\mathbf{r},\mathbf{r}')$ is defined as:

$$K(\mathbf{r},\mathbf{r}') = K(z - z') = \frac{1}{2\pi} \int_{-\pi}^{\pi} \frac{e^{-jk_n R}}{4\pi R} d\phi' = \frac{1}{4\pi^2} \int_{0}^{\pi} \frac{e^{-jk_n R}}{R} d\phi', \tag{69}$$

where $R = \sqrt{(z - z')^2 + 2r_0^2 - 2r_0^2 \cos\phi'}$. By letting $\alpha = \dfrac{\pi - \phi'}{2}$, the distance R can be rewritten as $R = R_{max}\sqrt{1 - \beta^2 \sin^2\alpha}$, where $R_{max} = \sqrt{(z - z')^2 + 4r_0^2}$, $\beta = 2\dfrac{r_0}{R_{max}}$; therefore, the kernel integral in (69) can be expressed as:

$$K(r,r') = \frac{1}{2\pi^2}\left[\frac{F(\beta)}{R_{max}} + \int_{0}^{\frac{\pi}{2}} \frac{e^{-jk_n R} - 1}{R} d\alpha\right], \tag{70}$$

where $F(\beta)$ is the complete elliptic integral of the first kind defined as:

$$F(\beta) = \int\limits_{0}^{\frac{\pi}{2}} \frac{1}{\sqrt{1 - \beta^2 \sin^2(\varsigma)}} d\varsigma, \tag{71}$$

The first term in (70) contains a logarithmic singularity as $\beta \to 1$, that is, $z \to z'$, while the second term is smooth and can be calculated with standard Gaussian quadrature formulae or approximated under the thin wire approximation as:

$$\int\limits_{0}^{\frac{\pi}{2}} \frac{e^{-jk_n R} - 1}{R} d\alpha \cong \frac{\pi}{2} \frac{e^{-jk_n \bar{R}} - 1}{\bar{R}}, \tag{72}$$

where $\bar{R} = \sqrt{(z - z')^2 + 2r_0^2}$.

When the first term in (70) is included in the source linear integral (68), the logarithmic singularity must be accurately integrated on the linear element by using the modified quadrature rules proposed by Wandzura in [48].

5.2 Sommerfeld Integrals

The numerical computation of the field quantities in the dynamic case requires the efficient evaluation of both the Fourier-Bessel integral transforms (commonly known as Sommerfeld integral), which are semi-infinite range integrals with oscillating Bessel function kernels.

The integrals that need to be computed are of the type:

$$G(\rho, z; \omega) = I = \int\limits_{0}^{\infty} \underbrace{\tilde{K}(z, x; \omega) J_n(\rho x)}_{f(x)} dx. \tag{73}$$

where $\tilde{K}(z, x; \omega)$ is supposed to be smooth with respect to x and to behave asymptotically as a power function $O(x^q)$, that is:

$$f(x) \sim \frac{e^{-\varsigma x}}{x^{\mu}} \left[C + O\left(\frac{1}{x}\right) \right], \tag{74}$$

where C is a constant and where ς and q can easily be determined. In (73), $J_n(\rho x)$ is the oscillating Bessel function of the first kind of order n. To facilitate the integration, the semi-infinite range in (74) is usually split into two parts:

$$I = \int\limits_{0}^{a} f(x) dx + \int\limits_{a}^{\infty} f(x) dx \tag{75}$$

and the first path segment is deformed into the first quadrant of the complex plane to avoid the guided-wave poles and branch points of the integrand; the second integral is referred to as the *Sommerfeld tail*, which must be accurately evaluated. The value of a is selected to ensure that the integrand of the remaining tail segment is free of singularities.

The integration of the Sommerfeld tail is a well-known numerical topic in computational electromagnetics (CEM) and the interested reader is referred to the seminal works of Michalski and Mosig [49–54]. Mainly, two different techniques are widely used: the Weighted Averages (WA) method and the Double Exponential (DE) integration technique.

The basic idea behind the WA method is to transform the semi-infinite integral into an infinite series, by dividing it into partial finite integrals, which are individually computed (Michalski identified the WA algorithm as an "integration-then-summation" procedure akin to the classic Euler transformation [49]):

$$I = \int_a^\infty f(\cdot)\mathrm{d}x = \sum_{n=0}^\infty \int_{x_n}^{x_{n+1}} f(\cdot)\mathrm{d}x = \sum_{n=0}^\infty I_n^0. \tag{76}$$

where the break points x_n of the integration intervals are usually selected as equidistant points according as $\rho(x_{n+1} - x_n) = \pi$, being $q = \dfrac{\pi}{\rho}$ the asymptotic half-period of the Bessel function, with $x_0 = a$. Other possible choices of break points include the (exact) zero crossings and extremum points of the Bessel function in which case the subinterval length $q_n = x_{n+1} - x_n$ varies with n. The WA method should be able to deal with both proper and improper integrals: a proper integral would result in a convergent series, while an improper integral would be transformed into a divergent series.

To compute correctly the integral, we can initially apply the procedure introduced in 1882 by O. Holder, called the *H-means*: we compute a sequence of *partial sums* from the terms I_n^0 of the original series as:

$$S_n^1 = \sum_{i=0}^n I_i^0. \tag{77}$$

Then, we recursively calculate the mean values as:

$$S_n^{k+1} = \frac{S_n^k + S_{n+1}^k}{2}. \tag{78}$$

At every step, the length of the series S_n^{k+1} is reduced by one, thus generating a triangular scheme:

$$
\begin{array}{ccccccc}
S_1^1 & S_2^1 & \dots & S_n^1 & \dots & S_N^1 \\
S_1^2 & \dots & S_n^2 & \dots & S_{N-1}^2 \\
\vdots \\
S_1^5 = S^*,
\end{array}
$$

(79)

We obtain the final result S^* when the scheme reduces to a single number. For oscillating series, when the method converges, it always converges towards the value of the original integral in the Abel sense.

The iterated H-means shows a slow convergence. To speed up the convergence and make the method effective on more complex integrals, the WA method originally proposed by Michalski [49] replaces the simple arithmetic mean by a weighted mean:

$$
S_n^{k+1} = \frac{W_n^k S_n^k + W_{n+1}^k S_{n+1}^k}{W_n^k + W_{n+1}^k} = \frac{S_n^k + \eta_n^k S_{n+1}^k}{1 + \eta_n^k},
$$

(80)

where

$$
\eta_n^k = e^{q\zeta} \left(\frac{x_{n+1}}{x_n}\right)^{\mu - \frac{1}{2} + 2k}.
$$

(81)

The method has been generalized in [51] by Mosig where the iterated simple weighted means have been replaced by a unique multiple weighted mean, leading to a *generalized WA algorithm* as

$$
S^* = \frac{\displaystyle\sum_{n=1}^{N} w_n S_n^1}{\displaystyle\sum_{n=1}^{N} w_n},
$$

(82)

where

$$
w_n = e^{q\zeta} \binom{N-1}{n-1} x_n^{N-2-\mu}.
$$

(83)

As previously discussed, another procedure is the DE integration technique that was originally introduced by H. Takahasi and M. Mori [48, 49] to compute integrals with singularities at their endpoints. Later, the procedure was extended by M. Mori [55–59], passing through T. Ooura [60, 61] and other authors [62]. The strength of the DE quadrature method relies on a suitable change of variables that can send the singular endpoints to infinity; the method also exhibits a double exponential decrease in the Jacobian of the transformation, which extinguishes the original singularities at infinity, no matter their type in the original integral.

The first finite integral in (75) must be calculated along a suitable path on the complex plane that joins the origin to a point a placed on the real axis, passing sufficiently far from the integrand singularities. The integral that can be recast in the general form:

$$I = \int_a^b f(x)\,dx,\tag{84}$$

can be computed by applying the variable transformation (also called *tanh-sinh* transformation)

$$x = \phi(\xi) = \underbrace{\frac{b-a}{2}}_{m}\tanh\left(\frac{\pi}{2}\sinh\xi\right) + \underbrace{\frac{b+a}{2}}_{q},\tag{85}$$

where $\phi(\xi)$ is an analytic function in $(-\infty,+\infty)$. It is straightforward to obtain the following resulting integral:

$$I = \int_{-\infty}^{+\infty} f[\phi(\xi)]\phi'(\xi)\,d\xi = \int_{-\infty}^{+\infty}\left[m\tanh\left(\frac{\pi}{2}\sinh\xi\right) + q\right]$$
$$m\frac{\frac{\pi}{2}\cosh\xi}{\cosh^2\left(\frac{\pi}{2}\sinh\xi\right)}\,d\xi.\tag{86}$$

The integral (86) can now be computed applying a standard trapezoidal approach to obtain the final DE formula (87)

$$I = hm\sum_{n=-N}^{+N}\left\{m\tanh\left[\frac{\pi}{2}\sinh(nh)\right] + q\right\}\underbrace{\frac{\frac{\pi}{2}\cosh(nh)}{\cosh^2\left[\frac{\pi}{2}\sinh(nh)\right]}}_{A_n},\tag{87}$$

where h is the amplitude and $2N$ is the number of the integration steps. Since the DE transformation ensures a DE decay of the integrand, the truncation at $2N + 1$ terms in the trapezoidal formula (87) is properly chosen to ensure convergence with a desired relative accuracy. Interestingly, the weights A_n of the DE formula can be computed in an efficient manner, avoiding to fall into the overflow of the denominator term through a recurrence relation $A_{n+1} = r_n A_n$, where

$$r_n = \frac{\cosh h + \sinh h \tanh(nh)}{\left[\cosh s_n + \phi(nh)\sinh s_n\right]^2} \tag{88}$$

and where $s_n = \pi \sinh \dfrac{h}{2} \cosh\left[\left(n + \dfrac{1}{2}\right)h\right]$.

An adaptive scheme can efficiently be obtained by using an automatic procedure, in which the integration step h is repetitively halved until the required relative precision is obtained. To this end, the DE formula has the practical advantage of an equidistant distribution of the sampling points, which allows the use of the values computed at previous steps to reduce the evaluations of the integrand function.

As to the computation of the Sommerfeld tail, the approach developed in Ooura and Mori [57] and Ogata [63] can be applied for the integration over the semi-infinite range $(0, \infty)$, as adapted in Polimeridis and Mosig [62] to the specific case, where the integration starts from a finite real point $a \geq 0$. Let the transformation be:

$$\rho x = \phi(\xi) = \underbrace{\frac{\pi}{h}\xi \tanh\left(\frac{\pi}{2}\sinh \xi\right)}_{\phi_1(\xi)} + \underbrace{a\rho \operatorname{sech}\left(\frac{\pi}{2}\sinh \xi\right)}_{\phi_2(\xi)} \tag{89}$$

where h is the step size that has been properly chosen to guarantee an efficient convergence. The original transformation proposed by Ogata [63] is the first term $\phi_1(\xi)$. The contribution of Polimeridis and Mosig [62] was to add the second term $\phi_2(\xi)$, whose role is to map the starting point of integration $x = a$ to $\xi = 0$, having suitably chosen this term to ensure that its effect rapidly vanishes as $\xi \to \infty$. In this way, it is possible to maintain the DE convergence of the original transformation.

Introducing the DE transformation, we readily obtain:

$$I = \int_a^\infty \tilde{K}(z, x; \omega) J_n(\rho x)\, dx = \int_0^\infty \underbrace{\frac{1}{\rho}\tilde{K}\left(z, \frac{\phi(\xi)}{\rho}; \omega\right) J_n\left[\phi(\xi)\right]\phi'(\xi)}_{\overset{0}{F_n}(\xi)}\, d\xi \tag{90}$$

where

$$\phi'(\xi) = \frac{\pi}{2h}\left\{\begin{array}{l}\left[\cosh\xi\,sech^2\left(\frac{1}{2}\pi\sinh\xi\right)\right]\left[\pi\xi - a\rho h\sinh\left(\frac{1}{2}\pi\sinh\xi\right)\right]\\ +2\tanh\left(\frac{1}{2}\pi\sinh\xi\right)\end{array}\right\} \quad (91)$$

After some algebraic manipulations, taking advantage of the odd nature of the integrand in (90) and introducing an appropriate quadrature formula based on the zeros of the Bessel functions as originally proposed in Ogata [63], it is possible to obtain the final DE quadrature formulae for the Bessel functions of order 0 and 1 as

$$I_0 = \int_a^\infty \tilde{K}(z,x;\omega)J_0(\rho x)dx = h\sum_{k=1}^N w_{0k}\tilde{F}_0\left(\bar{\xi}_{0k}\right)\phi'\left(\bar{\xi}_{0k}\right) \quad (92)$$

$$I_1 = \int_a^\infty \tilde{K}(z,x;\omega)J_1(\rho x)dx = h\sum_{k=1}^N w_{1k}\tilde{F}_1\left(\bar{\xi}_{1k}\right)\phi'\left(\bar{\xi}_{1k}\right)$$
$$+\left(2h - \frac{1}{2}a\rho h^2\right)\tilde{F}_1(a\rho) \quad (93)$$

with $\bar{\xi}_{nk} = \dfrac{h}{\pi}\chi_{nk}$, where χ_{nk} are the n-th zeros of the Bessel functions $J_0(\cdot)$ and $J_1(\cdot)$, respectively. Furthermore, the weights w_{nk} of the integration formulas are defined as:

$$w_{0,1k} = \frac{2}{\pi\chi_{0,1k}J_{1,2}^2\left(\chi_{0,1k}\right)}. \quad (94)$$

As $\xi \to \infty$, the quadrature nodes approach double exponentially to the zeros of the associated Bessel functions, allowing us to truncate the infinite sum in (92) and (93) at moderate N.

References

1 Liew, A.C. and Darveniza, M. (1974). Dynamic model of impulse characteristics of concentrated earths. *Proceedings of the Institution of Electrical Engineers* 121 (2): 123–135(12).

2 Junping Wang, A.C.L. and Darveniza, M. (2005). Extension of dynamic model of impulse behavior of concentrated grounds at high currents. *IEEE Transactions on Power Delivery* 20 (3): 2160–2165.

3 Ramamoorty, M., Narayanan, M.M.B., Parameswaran, S., and Mukhedkar, D. (1989). Transient performance of grounding grids. *IEEE Transactions on Power Delivery* 4 (4): 2053–2059.

4 Geri, A. (1999). Behaviour of grounding systems excited by high impulse currents: The model and its validation. *IEEE Transactions on Power Delivery* 14 (3): 1008–1017.

5 Devgan, S.S. and Whitehead, E.R. (1973). Analytical models for distributed grounding systems. *IEEE Transactions on Power Apparatus and Systems* PAS-92 (5): 1763–1770.

6 Verma, R. and Mukhedkar, D. (1980). Impulse impedance of buried ground wire. *IEEE Transactions on Power Apparatus and Systems* PAS-99 (5): 2003–2007.

7 Mazzetti, C. and Veca, G.M. (1983). Impulse behavior of ground electrodes. *IEEE Power Engineering Review* PER-3 (9): 46.

8 Velazquez, R. and Mukhedkar, D. (1984). Analytical modelling of grounding electrodes transient behavior. *IEEE Transactions on Power Apparatus and Systems* PAS-103 (6): 1314–1322.

9 Menter, F.E. and Grcev, L. (1994). EMTP-based model for grounding system analysis. *IEEE Transactions on Power Delivery* 9 (4): 1838–1849.

10 Heimbach, M. and Grcev, L.D. (1997). Grounding system analysis in transients programs applying electromagnetic field approach. *IEEE Power Engineering Review* 17 (1): 45–46.

11 Celli, G., Ghiani, E., and Pilo, F. (2017). Behaviour of grounding systems: A quasi-static EMTP model and its validation. *2010 30th International Conference on Lightning Protection, ICLP 2010* 85: 24–29.

12 Liu, Y., Theethayi, N., and Thottappillil, R. (2005). An engineering model for transient analysis of grounding system under lightning strikes: Nonuniform transmission-line approach. *IEEE Transactions on Power Delivery* 20 (2): 722–730.

13 Dawalibi, F. (1986). Electromagnetic fields generated by overhead and buried short conductors. Part 2 – Ground networks. *IEEE Power Engineering Review* PER-6 (10): 33–34.

14 Dawalibi, F. (1986). Electromagnetic fields generated by overhead and buried short conductors Part 2 – Ground networks. *IEEE Transactions on Power Delivery* 1 (4): 112–119. doi: 10.1109/TPWRD.1986.4308037.

15 Nekhoul, B., Guerin, C., Labie, P., Meunier, G., Feuillet, R., and Brunotte, X. (1995). A finite element method for calculating the electromagnetic fields generated by substation grounding systems. *IEEE Transactions on Magnetics* 31 (3): 2150–2153.

16 Grcev, L. and Dawalibi, F. (1990). An electromagnetic model for transients in grounding systems. *IEEE Transactions on Power Delivery* 5 (4): 1773–1781.

17 Meliopoulos, A.P. and Moharam, M.G. (1983). Transient analysis of grounding systems. *IEEE Power Engineering Review* PER-3 (2): 31.

18 Grcev, L.D. (1996). Computer analysis of transient voltages in large grounding systems. *IEEE Transactions on Power Delivery* 11 (2): 815–823.

19 Olsen, R.G. and Willis, M.C. (1996). A comparison of exact and quasi-static methods for evaluating grounding systems at high frequencies. *IEEE Transactions on Power Delivery* 11 (2): 1071–1078.

20 Ala, G., Francomano, E., Toscano, E., and Viola, F. (2004). Finite difference time domain simulation of soil ionization in grounding systems under lightning surge conditions. *Applied Numerical Analysis & Computational Mathematics* 1 (1): 90–103.

21 Otani, K., Baba, Y., Nagaoka, N., Ametani, A., and Itamoto, N. (2014). FDTD surge analysis of grounding electrodes considering soil ionization. *Electric Power Systems Research* 113: 171–179.

22 Li, J., Yuan, T., Yang, Q., Sima, W., Sun, C., and Meng, X. (2011). Finite element model of grounding system considering soil dynamic ionization. *Zhongguo Dianji Gongcheng Xuebao/Proceedings of the Chinese Society for Electrical Engineering* 31 (22): 149–157.

23 Sima, W., Zhu, B., Yuan, T., Yang, Q., and Wang, J. (2016). Finite-element model of the grounding electrode impulse characteristics in a complex soil structure based on geometric coordinate transformation. *IEEE Transactions on Power Delivery* 31 (1): 96–102.

24 Li, J., Yuan, T., Yang, Q., Sima, W., and Sun, C. (2010). Finite element modeling of the grounding system in consideration of soil nonlinear characteristic. *2010 International Conference on High Voltage Engineering and Application, ICHVE 2010*, 164–167.

25 Zhang, B., Cui, X., Member, S., Zhao, Z., He, J., and Li, L. (2005). Numerical analysis of the influence between large grounding grids and two-end grounded cables by the moment method coupled with circuit equations, 20 (2): 731–737.

26 Papalexopoulos, A.D. and Meliopoulos, A.P. (1987). Frequency dependent characteristics of grounding systems. *IEEE Power Engineering Review* PER-7 (10): 43–44.

27 Otero, A.F., Cidrás, J., and Del Álamo, J.L. (1999). Frequency-dependent grounding system calculation by means of a conventional nodal analysis technique. *IEEE Transactions on Power Delivery* 14 (3): 873–878.

28 Visacro, S. and Soares, A. (April 2005). HEM: A model for simulation of lightning-related engineering problems. *IEEE Transactions on Power Delivery* 20 (2): 1206–1208.

29 Ahmed, M.R. and Ishii, M. (2011). Electromagnetic analysis of lightning surge response of interconnected wind turbine grounding system. *2011 International Symposium on Lightning Protection, XI SIPDA 2011*, 226–231.

30 Ahmed, M.R. and Ishii, M. (2012). Electromagnetic analysis of interconnected groundings at wind farm. *IEEJ Transactions on Power and Energy* 132 (7): 684–689.

31 Li, Z.X. and Bin Fan, J. (2008). Numerical calculation of grounding system in low-frequency domain based on the boundary element method. *International Journal for Numerical Methods in Engineering* 73 (5): 685–705.

32 Li, Z.X. and Chen, W.J. (2007). Numerical simulation grounding system buried within horizontal multilayer earth in frequency domain. *Communications in Numerical Methods in Engineering* 23 (1): 11–27.

33 Li, Z.X., Li, G.F., Bin Fan, J., and Zhang, C.X. (2009). Numerical calculation of grounding system buried in vertical earth model in low frequency domain based on the boundary element method. *European Transactions on Electrical Power* 19 (8): 1177–1190.

34 Li, Z.X., Li, G.F., and Bin Fan, J. (2010). Numerical simulation of substation grounding system in low-frequency domain based on the boundary element method with linear basis function. *International Journal for Numerical Methods in Biomedical Engineering* 26 (12): 1899–1914.

35 Li, Z.X., Li, G.F., Fan, J.B., and Yin, Y. (2011). Quasi-static complex image method for a current point source in horizontally stratified multilayered earth. *Progress in Electromagnetics Research (PIER) B* 34: 187–204.

36 Li, Z.X., Yin, Y., Zhang, C.X., and Zhang, L.C. (2014). A mathematical model for the transient lightning response from grounding systems. *Progress in Electromagnetics Research (PIER) B*: 47–61. doi:10.2528/PIERB13101908.

37 Li, Z.X., Gao, K.L., Yin, Y., and Ge, D. (2014). Transient lightning responses of grounding systems buried in horizontal multilayered earth with a hybrid method. *Journal of Electrostatics* 72 (5): 381–386.

38 Li, Z.X., Yin, Y., Zhang, C.X., and Zhang, L.C. (2015). Numerical simulation of currents distribution along grounding grid buried in horizontal multi-layered earth in frequency domain based dynamic state complex image method. *Electric Power Components and Systems* 43 (14): 1573–1582.

39 Araneo, R., Lovat, G., and Celozzi, S. (2014). Transient response of grounding systems of wind turbines under lightning strikes. *Proceedings of the EMC Europe International Symposium*, 1080–1085.

40 Stracqualursi, E., Araneo, R., Burghignoli, P., Celozzi, S., and Lovat, G. (2018). Offshore wind towers interaction through their grounding systems. *IEEE International Symposium on Electromagnetic Compatibility*, vol. 2018–August, 908–912.

41 Wedepohl, L.M. and Wilcox, D.J. (1973). Transient analysis of underground power-transmission systems. System-model and wave-propagation characteristics. *Proceedings of the Institution of Electrical Engineers* 120 (2): 253–260.

42 Weng Cho, C. (1999). Waves and Fields in Inhomogenous Media. IEEE Press.

43 Michalski, K.A. and Zheng, D. (1990). Electromagnetic scattering and radiation by surfaces of arbitrary shape in layered media, part I: Theory. *IEEE Transactions on Antennas and Propagation* 38 (3): 335–344.

44 Michalski, K.A. and Mosig, J.R. (1997). Multilayered media green's functions in integral equation formulations. *IEEE Transactions on Antennas and Propagation* 45 (3): 508–519.

45 Kourkoulos, V.N. and Cangellaris, A.C. (2006). Accurate approximation of green's functions in planar stratified media in terms of a finite sum of spherical and cylindrical waves. *IEEE Transactions on Antennas and Propagation* 54 (5): 1568–1576.

46 Wilton, D.R. and Champagne, N.J. (2006). Evaluation and integration of the thin wire kernel. *IEEE Transactions on Antennas and Propagation* 54 (4): 1200–1206. doi: 10.1109/TAP.2005.872569.

47 Resende, U.C., Moreira, M.V., and Afonso, M.M. (2014). Evaluation of singular integral equation in MoM analysis of arbitrary thin wire structures. *IEEE Transactions on Magnetics* 50 (2): 457–460. doi: 10.1109/TMAG.2013.2281315.

48 Ma, J., Rokhlin, V., and Wandzura, S. (1996). Generalized Gaussian quadrature rules for systems of arbitrary functions. *SIAM Journal on Numerical Analysis* 33 (3): 971–996.

49 Michalski, K.A. and Member, S. (1998). *Extrapolation methods for sommerfeld integral tails.* 46 (10): 1405–1418. doi: 10.1109/8.725271.

50 Mosig, J.R. and Gardiol, F.E. (1983). Analytical and numerical techniques in the Green's function treatment of microstrip antennas and scatterers. *IEE Proceedings H Microwaves Optic Antennas* 130 (2): 175–182.

51 Mosig, J. (April 2012). The weighted averages algorithm revisited. *IEEE Transactions on Antennas and Propagation* 60 (4): 2011–2018. [Online]. Available: http://ieeexplore.ieee.org/lpdocs/epic03/wrapper.htm?arnumber=6143991.

52 Golubovic, R., Polimeridis, A.G., and Mosig, J.R. (2012). Efficient algorithms for computing sommerfeld integral tails. *IEEE Transactions on Antennas and Propagation* 60 (5): 2409–2417.

53 Golubović, R., Polimeridis, A.G., and Mosig, J.R. (2013). The weighted averages method for semi-infinite range integrals involving products of bessel functions. *IEEE Transactions on Antennas and Propagation* 61 (11): 5589–5596.

54 Mosig, J.R. (2013). Weighted averages and double exponential algorithms. *IEICE Transactions on Communications* E96-B (10): 2355–2363.

55 Mori, M. (2005). Discovery of the double exponential transformation and its developments. *Publications of the Research Institute for Mathematical Sciences* 41 (4): 897–935.

56 Ooura, T. and Mori, M. (1999). A robust double exponential formula for Fourier-type integrals. *Journal of Computational and Applied Mathematics* 112 (1–2): 229–241.

57 Takahasi, B.H. and Mori, M. (1974). Double exponential formulas for numerical integration 9: 721–741.

58 Mori, M. and Sugihara, M. (2001). The double-exponential transformation in numerical analysis. *Journal of Computational and Applied Mathematics* 127 (1–2): 287–296.

59 Mori, M. (1988). The double exponential formulas for numerical integration over the half infinite interval: 367–379.

60 Ooura, T. (2001). A continuous Euler transformation and its application to the Fourier transform of a slowly decaying function. *Journal of Computational and Applied Mathematics* 130 (1–2): 259–270.

61 Ooura, T. (2005). A double exponential formula for the Fourier transforms. *Publications of the Research Institute for Mathematical Sciences* 41 (4): 971–977.

62 Golubović Niciforović, R., Polimeridis, A.G., and Mosig, J.R. (2011). Fast computation of sommerfeld integral tails via direct integration based on double exponential-type quadrature formulas. *IEEE Transactions on Antennas and Propagation* 59 (2): 694–699.

63 Ogata, H. (2005). A numerical integration formula based on the bessel functions. *Publications of the Research Institute for Mathematical Sciences* 41 (4): 949–970.

Appendix 2

Cable Failures in Renewable Energy Systems

CONTENTS

---

*The effort to understand the Universe is one of the very few things
that lifts human life above the level of farce, and gives it some of the grace of
tragedy.*

Steven Weinberg

Electrical Safety Engineering of Renewable Energy Systems. First Edition. Rodolfo Araneo and
Massimo Mitolo.
© 2022 by The Institute of Electrical and Electronics Engineers, Inc. Published 2022 by
John Wiley & Sons, Inc.

1 Cable Failures in Renewable Energy Systems: Introduction

The repeated occurrence of damages to underground electrical, telephone, or network communications cables caused by subterranean termites is a well-known phenomenon (Beal et al. 1973; Beal & Bultman 1978; Lenz et al. 1992, 2012; Su & Scheffrahn 1998, 2010; Scheffrahn et al. 2015), mainly in regions where termites show their highest levels of biodiversity and species richness (Eggleton 2000; Carezer et al. 2020).

Renewable energy installations may employ high-voltage underground cables for the inter-connection of sources and the substation(s), with a total length in the order of tens of kilometers. Severe termite-caused damages to such cables, and consequent loss of energy production, have been recently reported worldwide in several sites (Figure 1).

IEEE Std 1264 "*Guide for Animal Deterrents for Electric Power Supply Substations*" does recognize that many animals, including insects, can affect the operation of electric systems.

Termites are wood-eater and wood-borer insects, which use wood, and other vegetal material containing cellulose, as food. Termites can digest and metabolize cellulose using a series of symbiotic micro-organisms present in their intestine. The species involved in the cable failures herein discussed is the *subterranean termites*, also known as *underground heartwood termites*, which are

Figure 1 MV cable damaged by termites.

frequently recognized as pests, mainly in wood plantations and other man-influenced habitats.

Termites may bite the external surface of the jacket, whose typical thickness may be 2.4 mm only, which can be completely punctured. Termites may then go through the copper wires of the concentric neutral and pierce the semi-conductive screen and the cable insulation (e.g., XLPE). Soil dust and moisture may then contaminate the cable and may trigger partial discharge phenomena, which can cause cable failure in days or weeks.

The insulation of cables is not a food source for termites; however, looking for potential ligneous substances, termites may attempt to bore, or taste, relatively soft materials, such as plastic substances, including polyethylene, and even some metals, such as copper. Underground cables can therefore be severely damaged, and phenomena of electric dispersion and loss of efficiency may occur, causing the need for repair and maintenance (Beal et al. 1973; Beal & Bultman 1978; Lenz et al. 1992, 2012).

Termites can damage the cable jacket but cannot penetrate a cable screen, if present; however, the screen becomes exposed to moisture, and its corrosion may occur. The screen corrosion causes premature aging of cables, and this initial mechanical damage may escalate into insulation deterioration. If the screen is not present in the cable assembly, after damaging the jacket, the termites may penetrate inside the cable interstices and come in contact with the cable filler, which contains cellulose; short circuits may then occur.

Mechanical damages are responsible for the majority of cable faults. Some damages may not cause instantaneous failures, but the conditions may deteriorate and cause a fault after several months or years. Cables are generally very durable, typically 40 to 50 years. Studies by Sandia National Laboratories, as well as by Japanese researchers, show that most cables rated for a 40-year service maintain their electrical functionality over 60 years of operation.

2 Possible Solutions

To prevent insect-caused damages, several solutions may be implemented. The underground cables may be equipped with an additional metallic sheath, made of iron or aluminum, which surrounds the outer plastic sheath and protects against the termites. Drawbacks are higher costs, but also possible electric induction phenomena, which may impact the ampacity.

Cables may be embedded in concrete or protected by a prefabricated concrete jacket, which is effective against the termites' attacks providing that each section is properly sealed, including the inspection wells.

Additional protective "hard" plastic sheaths (e.g., nylon) that surround the cables and their outer plastic sheath may be employed. However, cable damages produced by certain species of subterranean termites (including *Coptotermes*) may be reduced but hardly eliminated. Protective plastic sheets may contain biologically active substances that regulate the termites' reproduction, but with the risk of potential ecological damages to local environment and to the biodiversity.

Termite nests and mounds in the areas of the cable burial may be removed, which limits the termites' attacks for a few months to a year. This solution is based on the visual or remote detection of the termite mounds, which add costs to the installation, and may again impact the local environment and the biodiversity.

The massive use of pesticide would provide a short-term protection, but unacceptable ecological damages to ground and running waters may occur.

The cable could be buried at greater depths (e.g., 1.5–2 m), which would greatly decrease the chance of intersecting the feed paths of the termites; although effective, this solution would substantially increase the initial costs and the costs for cable maintenance and repair.

If the cable were installed above ground, the issue of the termites would not be present; however, the cable would be exposed to sunlight and atmospheric agents, as well as to possible vandalism and rodents.

2.1 Optimal Solutions

Some of the above solutions, although effective, may be cost-prohibitive. Alternatively, the authors propose the use of strips or tape made of fiberglass (or carbon fiber) to be wrapped around the cables to form an additional outer sheath, which would nearly completely reduce the likelihood of termite attacks. A similar solution already available in the market is used for the protection against rodents of underground fiber optics cables, although its application against termites has not yet been verified.

An effective alternative solution may likely be a combination of relatively low-cost differentiated actions, as follows:

1. Accurate identification of the points where cable damages occurred along the entire path of the underground cables.
2. Environmental analysis of the above points to verify the presence of trees, trunks, stumps, or clusters of roots, within a radius of about 100 m.
3. Mechanical removal of all possible ligneous material and termite nests around the above-identified points, as well as of man-made subterranean cavities, to prevent further damages and avoid further termites' colonization.

4. Protection with a layer of fiberglass tape the sections of underground cables closest to forested areas and trees, or isolated stumps.
5. Greater burial depths of the underground cables (e.g., 1.5–2 m), at least the runs closer to forested areas and trees, or isolated stumps.

2.2 Termite Attacks Prevention

A preliminary investigation of the cable burial site should always be carried to ascertain the possible or likely presence of termites in the subsoil. Due to the nearly worldwide distribution of termites, attacks are more likely to occur in areas with a higher termite biodiversity, such as the tropical and subtropical South America, southern North America and Central America, southern Asia and the whole Tropical and subtropical Asia, Northern Australia, Southern, and Eastern Africa.

Once the actual or probable presence of termites (especially subterranean termites) has been determined in a particular site, electrical cables should not be buried within 100 m of forested areas, isolated trees, partially buried stumps, and wooden artifacts. Cables should be buried at a depth of 1.5–2 m and/or protected by wraps made of fiberglass or carbon fiber.

3 Non-destructive Methods for Cable Testing and Fault-locating

Fault location techniques for underground cable systems have been applied since the 1800s. Since then, the use of cables has dramatically increased, and the technologies for locating cable faults have been greatly improved. Sheath and cable diagnostics can make operators aware of latent defects within the cable network and provide information on whether a cable system is safe and ready to operate.

Cable failure, causing faults within the renewable energy system, may be caused by mechanical damages, insulation aging and deterioration, overvoltages, overloads, moisture retention in the insulation, etc. Faults can be subdivided into two types: *series faults* and *parallel faults*. The series fault occurs when a gap (i.e., an open) occurs between two or more points within the same conductor of a cable. The parallel fault occurs when the insulation resistance between wires at different voltages drops down to the point that the normal operating voltage can no longer be withstood. According to the fault type, common faults are ground, short circuit, and open circuit. The termites cause mechanical damage, which gives rise to parallel faults.

One of the oldest and most popular fault-locating techniques for cables is the capacitor discharge technique or "thumping" the cable. A high-voltage surge wave generator (the "thumper") creates a surge of energy that travels along the cable conductor. If it reaches an area of dielectric breakdown or of a failure of the insulating material, the energy will discharge through the gap of the insulation. The impressed fault current will then return through the neutral or the shield of the cable to the surge wave generator, where it will be safely dissipated into the earth. The discharged energy will also determine a small explosion, which will cause a percussion and a sound wave to travel up through the layers of the soil. As a result, a thumping sound will be heard on the soil surface. The repair crew walks along the soil, listening for the thumping sound and can pinpoint the location of the fault. The soil is then removed, and repairs to the deenergized cable can be performed. This technique is very time consuming and potentially harmful if misused since it could lead to partial destructions of the cable.

The importance of non-destructive tests for locating faults is properly discussed in IEEE Std. 1234-2019 "*Guide for Fault-Locating Techniques on Shielded Power Cable Systems*," where it is stated: "proper locating of high resistance or intermittent cable faults, which are the majority of the faults on cables with extruded dielectric insulation, is considered tedious, inconsistent, and time consuming. Therefore, reclosing, re-fusing, burning, and thumping at unnecessarily high voltage and energy levels, in order to generate an open or short circuit, are frequently used without consideration of cable and equipment properties. The danger of activating dormant faults, generating new faults, or damaging utility and customer equipment by improper locating methods is not always recognized. By establishing cable fault-locating guidelines and training programs that incorporate recommended cable fault-locating measurements and techniques, cable owners can realize substantial savings in labor, cable, and equipment replacement, and avoid unnecessary losses from customer outages. during installation, mechanical tension is too large and damage cables; and damaged cables by excessive bending."

Current technologies can monitor the status of the entire cable system, including connections, splices, and terminations, even though the primary focus is the electrical insulation and the integrity of the jacket. The majority of conductor problems can be identified and pinpointed through impedance measurements and electronic tests. However, insulation problems are generally difficult to identify and locate. While conductor issues can be identified through several in situ tests, there is only one family of in situ testing techniques for checking the cable insulation, which is referred to as *electrical testing*.

In situ electrical tests allow the assessment of cables while they remain in operation. Among the most common types of electrical tests (also included in IEEE Std. 1234-2019) are:

1) The *insulation resistance test*, which measures the leakage current through the degraded cable insulation.
2) *Impedance measurement tests*, including the LCR test, later on, detailed, which performs lumped parameter measurements of inductance or capacitance, ac resistance, and dc resistance along the cables, including connectors, splices, and other components.
3) *Reflectometry* or *"cable radar" tests*, such as time domain reflectometry (TDR) or frequency domain reflectometry (FDR), which send a test signal through the cable and measure its reflection to determine the distance to the fault. Electrical methods such as insulation resistance, TDR, FDR, and LCR tests provide an overall picture of the cable's health, as well as an indication to perform any repairs that may be needed. It should be noted that most of these electrical tests are only possible if baseline data are available, so that significant change from the baseline, indicating the effects of dormant or high-resistance faults, may be recognized. If baseline data are not available, the characteristics of cables from similar installations may serve as a *de facto* baseline.

3.1 Insulation Resistance (IR) Test

The basic test for monitoring cable aging is to measure the insulation level of the cable, the connectors, and the end device, by using the insulation resistance test, which determines the leakage current from the degraded cable. The insulation resistance, which decreases as the temperature increases as later on detailed, is the resistance to the leakage current through and over the surface of the cable. Cables require two IR measurements: cable insulation and cable jacket insulation.

3.1.1 IR Measurement of the Cable Insulation (XLPE)

The insulation (i.e., XLPE, cross-linked polyethylene) measurement is performed between the core conductor and the concentric neutral and is executed testing the cable on both ends by connecting the megohmmeter positive lead to the concentric neutral and the negative to the conductor. The applied voltage must be not less than 10 kV d.c. International standardization organizations have not yet published standards on d.c. IR measurement of cables featuring XLPE insulation. However, in these last years, several utilities performed insulation assessment of medium-voltage cables by employing the methods included in IEEE Std. 43-2013 "IEEE Recommended Practice for Testing Insulation Resistance of Rotating Machinery," which take into account the effects of the temperature on the results of the IR measurement. This standard defines the insulation resistance measurement as: "the capability of the electrical insulation of a winding to resist the direct current. The quotient of applied direct voltage of negative polarity divided by current across machine insulation, corrected to 40°C and taken at a specified time (t) from start of voltage application. The voltage application time is usually 1 min. (IR_1) or 10 min. (IR_{10}); however, other values can be used. Unit conventions: values of 1 through 10 are assumed to be in minutes, values of 15 and greater are assumed to be in seconds."

3.1.2 IR Measurement of the Polyethylene (PE) Cable Jacket

The insulation resistance measurement of the polyethylene (PE) jacket is performed by applying a voltage between the concentric neutral and the cable installation environment and is executed testing the cable on both ends, by connecting the megohmmeter positive lead to the plant grounding system and the negative to the concentric neutral. The applied voltage must not be greater than 5 kV d.c., and the IR value of the PE jacket typically ranges between 100 and 300 MΩ/km at 20°C. The insulation resistance reading is evaluated individually and then compared to the readings of the other two cables composing the three-phase line under test. If the measured values are much smaller than the aforementioned typical values, the cable must be tested with a high-voltage bridge, as later on discussed.

3.2 High-Potential Test

The high-potential test applies a higher a.c. or d.c. voltages (e.g., twice the cable's rated voltage) between the conductors and the ground, or between the conductors and the cable shield. The voltage is typically applied for a specified time, such as 1 s, 1 min, or 5 min, with the intent of causing defective cables to fail when subjected to the test charge. Typically, a high-potential breakdown voltage and a high-potential withstand voltage are measured. The high-potential breakdown voltage is the point at which the cable insulation material begins to break down, and the high-potential withstand voltage is the highest voltage that can be applied before breakdown begins to occur. The results of these voltage measurements correlate with the condition of the cable insulation.

The high-potential test is a global circuit-wide in situ acceptance test. Though simple and effective, the a.c. high potentials can progressively weaken the cable insulation over multiple tests. In addition, both the a.c. and d.c. high-potential tests require the cable disconnection.

3.3 LCR Test

The LCR (i.e., inductance L, capacitance C, and resistance R) test is based on impedance measurements along the cable at specific frequencies to assess the characteristics of the conductor, insulating material, and end device. Imbalances, mismatches, or unexpectedly high or low impedances between the cable leads may indicate issues caused by cable degradation and aging, faulty connections and splices, or physical damages; for example, abnormal capacitance readings indicate a change in the insulation of the cable. LCR measurements can also identify circuit problems: for example, LCR readings can help detect problems such as moisture in the insulation or loose connections.

LCR and insulation resistance tests also allow the evaluation of other parameters that identify cable conditions, such as *insulation resistance*, *dielectric absorption ratio*, *polarization index*, *quality factor*, and *dissipation factor*.

3.3.1 Insulation Resistance (IR)

Insulation Resistance (IR) is the resistance to flow of a direct current through a dielectric, and it depends on the temperature according to the following formula:

$$IR_t = IR_{20} \ e^{-\alpha t}, \tag{1}$$

where IR_{20} is the insulation resistance value at 20°C (Ω), t is the temperature (°C), and α is the resistance temperature coefficient of the dielectric (°C^{-1}).

IR is also influenced by humidity level according to the degree of contamination of the insulating surfaces.

3.3.2 Dielectric Absorption Ratio (DAR)

Insulation resistance readings may change depending on how long the measurement voltage is applied (e.g., 60 s, etc.). Over time, a good insulation will slightly raise its level, whereas degraded insulation will not. DAR is an index of a cable's insulation quality over time and is a parameter that is independent of the temperature. Just like in the case of the insulation resistance measurement, DAR measures the leakage current. To determine DAR, the insulation resistance is measured 60 s after applying the test voltage, and that reading is then divided by the insulation resistance measurement after 30 s:

$$DAR = \frac{IR_{60}}{IR_{30}}. \tag{2}$$

DAR values of <1.0 usually indicate degradation in the insulating material, which may be due to dirt, moisture, cracking, aging, or other problems. DAR is somewhat subjective and should be considered in the context of insulation resistance; it is not an absolute indicator of insulation quality.

3.3.3 Polarization Index (PI)

The Polarization Index captures the change over time of the insulation resistance as a function of how long the measurement voltage is applied. Like DAR, PI measures the leakage current and is the ratio of two IR measurements. To calculate the PI, a measurement of the IR over a 10-min period is made, and a second one is made over a 1-min period.

The PI is then calculated as follows:

$$PI = \frac{IR_{10}}{IR_1}. \tag{3}$$

PI is defined in IEEE Standard 43-2000 as: "the quotient of the insulation resistance at time (t_2) divided by the insulation resistance at time (t_1). If times t_2 and t_1 are not specified, they are assumed to be 10 min and 1 min, respectively. Unit conventions: values of 1 through 10 are assumed to be in minutes, values of 15 and greater are assumed to be in seconds."

IEEE 43-2000 recommends a minimum PI value of 2.0 for most insulation systems. Lower readings may indicate moisture, contamination, embrittlement, or damage to the insulation system. IEEE 43-2000 also states: "when the IR_1 is higher than 5,000 MΩ, the P.I. may or may not be an indication of the insulation condition and is therefore not recommended as an assessment tool." This statement has caused the misconception that if the insulation resistance is greater than 5,000 MΩ, the insulation system is optimal.

3.3.4 Quality Factor (Q)

The Quality Factor (Q) applies to circuits with resistance, inductance, and capacitance and is the ratio of the energy stored to the energy dissipated in a system at a specific frequency. Therefore, the Q factor represents the cable's deviation from an "ideal" inductor.

3.3.5 Dissipation Factor (DF)

The *Dissipation Factor* is the ratio of the energy loss in a dielectric to the total energy transmitted through it; it represents a cable's deviation from an "ideal" capacitor. If the cable is free of defects or contaminants, its dielectric properties are similar to those of an ideal capacitor. If the cable dielectric contains impurities, the resistance of the insulation decreases, and it no longer acts as a perfect capacitor. The dissipation factor (also referred to as *Tan Delta*) is calculated as follows:

$$DF = \frac{1}{Q}. \tag{4}$$

The dissipation factor test is one of two related tests for measuring dielectric loss in a cable (the *power factor* test being the other). During this test, a steady-state ac voltage is applied between the conductor and the ground to stress the cable's insulation; a current composed of a charging current (indicating the cable insulation's capacitance) and a leakage current will flow. The higher the leakage current is, the weaker is the insulation. The dissipation factor test provides repeatable and quantifiable results on both accessible and partially inaccessible cables; however, the cable must be disconnected, and the risk of damages is present.

3.3.6 Time Domain Reflectometry (TDR) Test

The *Time Domain Reflectometry* (TDR) is the most popular cable-testing technique nowadays. The TDR test, is a simple yet powerful diagnostics tool first developed in the late 1940s. TDR is employed to locate defects along the cable, in a connector, or at passive devices at the end of the run, by sending a test signal through the conductors and measuring its reflection. In recent years, LCR measurements have been added to the TDR test to improve cable diagnostics and to help identify the nature and location of a fault along a cable.

The simplest and perhaps most important application of TDR is to locate an open or a short, moisture, or problems, such as an erratic behavior along a cable or at end devices. While it is always possible to tell if a circuit is open or short-circuited by measuring its loop resistance, the exact location of the fault can only be identified with a TDR, which is the method's major strength. In addition to being a nondestructive test, the TDR test also provides quantitative, trendable data with a relatively inexpensive equipment. Its disadvantages include the amount of expertise required to accurately read a TDR signature and the necessary disconnection of the cable at one end.

The TDR test involves sending an electrical step signal, or a pulse, through the cable to be tested and measuring its reflection, first over time and then converted into distance, to identify the location of any impedance discontinuity, or change, in the cable or at the end device (e.g., a load). Reflected voltage waves occur when the transmitted signal encounters an impedance mismatch or discontinuity (i.e., faults) in the cable, connector, or end device. Any such change in impedance along the cable due to a short, open, or other electrical condition can thus be identified and located. An increase in the reflected wave is indicative of an increase in impedance, and a decrease in the reflected wave is indicative of a decrease in impedance. Thus, the peaks and dips in a TDR plot are used to identify the location of normal and abnormal effects throughout the cable.

Novel measurement methods include frequency domain reflectometry (FDR) and line impedance resonance analysis (LIRA).

3.3.7 Arc Reflection (ARC) Test

The technique of Arc Reflection (ARC) combines the surge wave generator (i.e., the "thumper") and the pulse echo technique (TDR) for the purpose of locating high resistance shunt faults and intermittent shunt faults. Although the pulse echo technique is a very effective tool for a low-voltage cables, a problem occurs when it is used on medium-voltage cables. To locate vented trees, which form a hole in the cable insulation, a breakdown voltage is typically required. The pulse echo tool is very low voltage in nature, and this may cause the pulse echo energy to travel straight past the fault. If a high voltage breakdown surge is also simultaneously applied with the pulse echo pulses, when the fault breaks down due to the high voltage surge and the arc jumps through the faulty insulation, the pulse echo wave will reflect off that arc. The ARC method requires a coupling device (e.g., a filter or a power separation unit) into the circuit to allow the high-voltage surge generator and the low-voltage time domain reflectometer to be concurrently applied to the faulted cable. The surge momentarily reduces the resistance of the shunt fault, which creates a point of reflection for the low-voltage pulse transmitted by the time domain reflectometer and helps identify the fault.

The major feature of the ARC is that it helps locate any fault that will sustain an arc during the thumper pulse. Even though a surge wave generator is used, this method significantly reduces the amount of electrical stress that is associated with the thumping technique because only one or two surges are normally required for the pulse echo to pinpoint the fault, display that image and measure the distance.

One disadvantage of this method is that long cables with very highly attenuated dielectrics, or corrosion on the neutrals, can very quickly absorb the reflected pulses. Thus, the TDR should have enough transmitted pulse energy to overcome excessive cable length, or high corrosion, which would otherwise "bleed" down the signal. This may make it impossible to "see" the end of the cable or the actual fault, just because the signal attenuates before it can return to the receiver.

3.3.8 Bridge Methods

Bridge methods are among the earliest techniques of cable-fault locating. The best-known fault-locating bridge is probably the *Murray loop* (Figure 2).

To use a bridge fault-locating method, fault resistance and electrical continuity should be measured. By the aid of an insulation resistance tester, the conductor-to-metallic sheath (ground) or the conductor-to-conductor resistance is measured. If this resistance is in the hundreds of megohms, the fault should be conditioned with a burn-set to lower the fault resistance, preferably in the ohm or low kilohm range. With an ohmmeter, the resistance of the loop composed of the faulted conductor and an unfaulted conductor connected at the far end is measured.

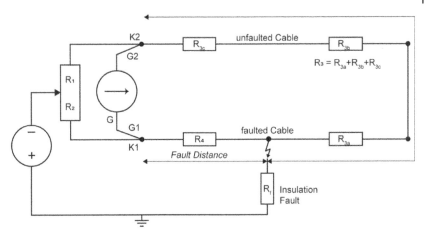

Figure 2 Circuit diagram for Murray bridge method.

If cable continuity and a low fault resistance exist, a dc bridge can be used to measure the distance to the fault. If the continuity test shows an open circuit, a TDR should be used to locate the fault.

The Murray bridge measures the distance to a low-resistance fault by joining one unfaulted conductor with a faulted conductor, applying a dc voltage to the conductors, and adjusting two variable resistors until a galvanometer placed across the joined conductors is nulled. From the known cable lengths and a ratio of adjusted variable resistors, the distance to the fault can be calculated. The Murray bridge method can be applied to conductor-to-ground (metallic shield) faults, conductor-to-conductor faults, and jacket faults (metallic shield-to-ground faults). The method may be used where other TDR-based methods are limited, such as for jacket faults, multi-branched cables, cables with corroded neutral, and cables with tape shield.

According to Figure 2, the bridge is balanced when the following equation is satisfied:

$$\frac{R_1}{R_{3a} + R_{3b} + R_{3c}} = \frac{R_2}{R_4} \tag{5}$$

Even though modern fault-locating bridges are microprocessor based, which calculate and display the distance to the fault in feet or as a percentage of the total cable length, it should be understood that this method is time consuming. Special attention should be given to these factors: (a) all bridge methods require at least one unfaulted conductor in addition to the faulted cable, unless the measurement can be performed on both ends simultaneously (e.g., a cable on a reel); (b) access to both cable ends is required;

(c) contact resistances and connecting wire resistance should be much less than the conductor resistance; (d) variations in the resistance of the faulted conductor should be considered; (e) stray dc and ac currents in the ground and in the cable will affect the measurement; (f) an unstable fault resistance will affect the measurement; (g) the actual conductor length, not the "point-to-point above ground" cable length, should be known; (h) multiple faults will affect the measurement; (i) the fault needs to be resistive (i.e., not a spark-gap type of fault.)

Other bridge methods cited in IEEE Std 1234-2019 are the Glaser bridge method and the voltage drop method.

With voltage drop methods, instead of balancing the bridge, a constant current is fed into the branches of the bridge and the voltage drop, over the particular section of the faulty wire, is measured. The first voltage-drop measurement is obtained by feeding a current at one end of the cable (i.e., the near end), followed by a second voltage-drop measurement from the remote (far) cable end. For this procedure, two unfaulted auxiliary conductors are needed. By using this four-point set-up, the influence from the jumper at the remote end and connection leads is reduced.

In the case of dc systems, only two conductors are available, and a four-point measurement is not possible, and a different procedure is used. With reference to Figure 3, the constant current is fed only from one end and R_N is obtained by the readings of V_N and I_N.

The distance-to-fault can be calculated by R_N by knowing the cable conductor resistance, corrected for the actual temperature. A second measurement for confirmation of the fault distance from the remote cable end is recommended. Calculation of distance-to-fault is obtained from:

$$L_N = L \frac{R_N}{R_{\text{TOTAL}}} \tag{6}$$

The pre-location of jacket faults can also be found by using the voltage drop measurement method. Figure 4 shows the connection diagram using two main conductors as auxiliary conductors.

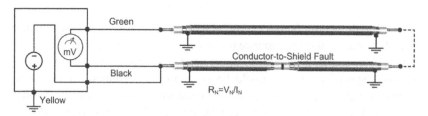

Figure 3 Connection of three-point voltage drop measurement for dc systems.

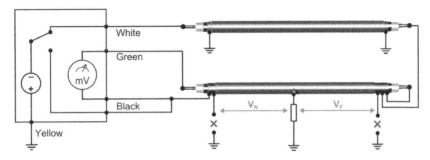

Figure 4 Voltage-drop method connection diagram for jacket fault locating.

For the first voltage-drop measurement, current is fed to the shield of the faulty cable at the near end, followed by a second voltage drop measurement from the remote (far) cable end. This four-point set-up method allows a reduction of the influence from the jumper at the remote end and connection leads. Calculation of distance-to-jacket-fault is obtained from:

$$L_N = L \frac{V_N}{V_N + V_F} \tag{7}$$

In the case, the main conductors cannot be used as auxiliary conductors (e.g., HV cable systems), the shield of a parallel cable can be instead used.

3.4 Cable Fault Analysis

The objective of the cable fault analysis is to identify the type of fault and determine its location through the selection of methods and voltages.
    Preferred methods are:

(a) Insulation resistance test to determine the faulty phase and the type of fault.
(b) High voltage withstand test for testing the electric strength of the cable insulation.
(c) Cable sheath testing (with LCR) to determine external cable damage (i.e., sheath faults).

3.4.1 Prelocation

The objective of the *prelocation* is to determine the fault location as precisely as possible to keep the subsequent pinpointing activities as brief and efficient as possible.
    Preferred methods are:

a) Time domain reflectometry for locating low-resistive faults and cable breaks, and for determining the cable length.

b) Arc reflection test, which is one of the most well-established and precise cable fault pre-location methods. High-resistive faults and breakdown faults are ignited by a single HV pulse and the fault distance is measured very precisely several times via the TDR technology and automatically analyzed.

c) Bridge methods.

3.4.2 Pinpointing

Pinpointing techniques consist of precisely locating the exact fault location for excavation and repair. These techniques employ audio (tone) frequency tracing and step voltage techniques.

Digital ground fault locator can pinpoint the cable fault by using the current field that is generated in the ground by the leakage conductive current flowing through the fault itself. The principle of operation is the step voltage technique. Data sheets of typical pinpointing instruments read: "Pinpointing means precisely locating faults in the cable sheath. These faults cause the measuring current to flow into the ground. When the measuring current exits the cable at the fault point, the current builds a voltage gradient which can be measured by earth rods and earth fault locators. The step voltage method accurately locates sheath faults: as it approaches the fault point, the step voltage potential increases. After it passes the fault, the potential decreases with reversed polarity. The change in polarity allows the fault to be located precisely."

The pinpointing is as precise as the prelocation method, but it cannot detect or recognize deviations of the cable route in the ground.

Preferred methods are:

a) *Acoustic pinpointing*: it is the most common method used to precisely locate high-resistive and flashover faults. High voltage pulses create electromagnetic pulses along the way to the fault and generate a flashover with an acoustically noticeable bang.

b) *Step voltage method*: used for the precise location of cable sheath faults. A voltage drop is generated at the fault, which can be located using ground electrodes and a receiver.

4 Sheath and Jacket Repairs

Termite-related downtime may range from momentary disturbances to long-term outages: large portions of PV installations may be tripped offline, including transmission circuits. In addition to the cost of the repairs, revenue is lost until the service is restored. Outages, sustained or momentary, may reduce the long-term confidence of all customers and become a critical factor for users. Most termite-related faults start as phase-to-ground faults, which may escalate to phase-to-phase faults. The analysis of the fault should identify the best course of action: immediately repair, defer repairs until a more convenient time, or plan to replace existing cables and lay new ones.

When the fault is precisely located and is limited to the outer sheath, the best option may be the timely repair, to put in service the distribution cable and restore the out-of-service section of the installation.

A possible solution may be the use of *wrap-around heat shrinkable insulating tubing* or *tape*, which are mainly used for repairing outer and inner sheaths of cables. They are also used for protection against the corrosion of metal parts of cables that are exposed to polluted environment.

Heat-shrinkable, polymeric, insulating materials have been used in outdoor environments for more than fifty years. The components are either molded or extruded forms of specialized, track-resistant compounds. The base polymer in the compounds must be semi-crystalline in order to function as a heat-shrinkable tape or tubing; the compound must contain a component that prevents tracking and must be capable of providing a hydrophobic surface throughout its service life.

Heat-shrinkable materials are made of semi-crystalline polymers that have been compounded with important additives. The components may be *cross-linked* (cured) or *non-cross-linked* (thermoplastic). Both polyethylene and polyvinyl chloride have been used in non-cross-linked, heat-shrinkable components. Cross-linking adds a very important feature to the polymer: it will not melt when heated. At normal operation temperatures, the mechanical strength of such polymer is provided by the presence of a semi-crystalline structure.

All heat-shrinkable components for electrical applications are semi-crystalline, and almost all are cross-linked. Cross-linking is a process that connects the long-chain, polymer molecules, so that they are unable to melt at temperatures above the crystalline melting temperature T_m.

The behavior of heat-shrinks against termites has not so far been investigated, therefore, the conservative working assumption is that they do not provide any additional protection against the insects.

Commercially available wrap-around sleeves are made from thermally stabilized, cross-linked, weather resistant, polymeric material, which is halogen free. The sleeves are coated internally with hot melt adhesive. The outer surface of the sleeves has thermochromic paint.

5 Termite Baiting Stations and Monitoring

Termite baiting stations, enjoying an increased popularity in the last 10 years, feature a slow-acting insecticide, which is consumed by termites and spread around the colony via trophallaxis, that is, the mutual exchange between social insects. Relatively small quantities of insecticide are deployed, and any remaining bait can be removed. The bait is slow acting, and the termites have to first find it and assume it, which requires extended monitoring. Current practice is for bait stations to be installed around the property by pest management professionals, who can periodically visit the stations, between 4 and 12 times per year, to monitor the termite activity. If termites are present, the bait is replaced with new; clearly, an online automated monitoring system would be cheaper and more effective.

The conventional termite monitoring has been based on the physical inspection of the installation. However, the cryptic nature of the subterranean termite and the possibility of their hidden presence require even experienced inspectors to use a range of detection devices, which basically fall into two groups: devices that directly detect the presence of termites, and devices that indirectly detect either the conditions that are favorable to termites, or the conditions created by the presence of termites.

Indirect devices include moisture meters, which are used to locate the termite favorable environment in structures. Termites require a moisture source and tend to be present in areas of higher moisture content. The presence of termites can also increase moisture levels. An inspector will therefore seek areas of unexpected moisture content, which will guide further investigations. Thermal imaging cameras can also be used to identify areas that may have increased moisture levels due to termite activity.

The presence of termites in a substantially enclosed space will lead to an increase in the concentration of carbon dioxide. Small probes that can be inserted into such spaces (e.g., a wall void in a substation) to sample the carbon dioxide concentration are also used to indicate the likely presence of termites.

Other systems detect elevated methane levels, which are also indicative of the presence of termites.

Termites may emit acoustic emissions, which can be detected by appropriate detectors. "Sniffer" dogs are also trained to detect the presence of termites with a 95.9% accuracy rate.

Once a renewable energy plant has been inspected, and remedial action was taken as necessary, a termite monitoring phase may be initiated. Generally, cellulose-based monitoring baits are placed in plastic bait stations around the property at intervals of approximately 3 m. Every one to three months, the monitor baits are inspected for the presence of termites, by a technician visiting the property, physically removing each cellulose bait and looking for the presence of termites. This is a time-consuming process, and more innovative bait systems have been developed to speed up the monitoring process. Some systems produce a visual change at the bait station when termites are present. Other systems rely on spring-loaded sections of the bait station to "pop up" when termites have eaten through a cellulose retaining mechanism, facilitating the assessment of the presence of termites.

Another system relies on the termites eating through electrically conductive tracks within the cellulose bait. The resulting change in electrical conductance can be recorded with a handheld device that is connected to terminals at the bait station.

6 Termite-proof Cables

Science and engineering have been applied to the development of protection strategies against termites. In early years, attempts to mitigate against termite attack involved backfilling around cables with soil containing insecticides such as DDT, Lindane, Aldrin, and Dieldrin. This practice has been banned in many countries due to the harmful effects on the environment and workers.

Collaborative research between research institutions and the cable industry identified in the Nylon-12 the more resistant insulating material against termite attack. As a result, extrusion of a Nylon-12 jacket over polyethylene (PE) cable-sheathing has become the industry standard in some countries (e.g., Australia) (Figure 5).

Cable sheathing compounds containing insecticides (e.g., Sevin, Lindane) were manufactured as an alternative to nylon; however, even though these chemicals were effective in fighting termites, it was found that continued attacks by follow-on members of the colony ultimately caused premature cable failure. Research concluded that Nylon-12 has unique blends of properties such as hardness, smooth surface finish, and flexibility, optimal for cable sheathing, which cannot be obtained with any other plastic materials with comparable termite resistance.

Figure 5 Nylon Cable.

Figure 6 Anti-termite resistant cable equipped with aluminum screen.

In the paper *"Resistance of polyamide and polyethylene cable sheathings to termites in Australia, Thailand, USA, Malaysia and Japan: A comparison of four field assessment methods,[1]"* the authors compare the behavior of cables against the attack of different species of termites. The authors concluded that termite-resistance insulating materials should always be evaluated in reference to the exact species of termites present in a specific country, and therefore it should be localized; plastic materials based on successful trials with species outside a country may not, in fact, be effective in another.

Special cables featuring a metallic barrier, such as brass, copper, aluminum, and stainless-steel tapes, may also be used against termite attacks (Figure 6).

Normally, two helically applied tapes, the upper one overlapping the gap in the lower one, may be incorporated into the cable design. In the case of armored cables, the brass tapes may be applied under the bedding of the armor. For unarmored cable, the brass tapes can be applied over the normal PVC or other extruded sheath followed by a PVC sheath over the brass tapes.

While the metallic barrier is effective against termites, further installation requirements become necessary to ensure that the metallic barrier is brought to ground potential at the termination(s). It is also important to ensure that bend radii are respected to prevent unwanted openings in the armor, which may give the termites access to the inner cable components. All this adds additional time and cost compared to the polymeric-protected cables.

1 International Biodeterioration & Biodegradation, 66 (2012) pp. 53–62.

Index

Electrical Safety Engineering of Renewable Energy Systems. First Edition. Rodolfo Araneo and Massimo Mitolo.
© 2022 by The Institute of Electrical and Electronics Engineers, Inc. Published 2022 by John Wiley & Sons, Inc.

Printed and bound by CPI Group (UK) Ltd, Croydon, CR0 4YY

16/04/2025

14658419-0001